Young People, Wellbeing and Placemaking in the Arctic

Youth are usually not (yet) decision-makers in politics or in business corporations, but the sustainability of Arctic settlements depends on whether or not youth envision such places as offering opportunities for a good future. This is the first multidisciplinary volume presenting original research on Arctic youth.

This edited book presents the results of two research projects on youth wellbeing and senses of place in the Arctic region. The contributions are united by their focus on agency. Rather than seeing youth as vulnerable and possible victims of decisions by others, they illustrate the diverse avenues that youth pursue to achieve a good life in the Arctic. The contributions also show which social, economic, political and legal conditions provide the best frame for youth agency in Arctic settlements.

Rather than portraying the Arctic as a resource frontier, a hotspot for climate change and a place where biodiversity and traditional Indigenous cultures are under threat, the book introduces the Arctic as a place for opportunities, the realization of life trajectories and young people's images of home. Rooted in anthropology, the chapters also feature contributions from the fields of sociology, geography, sustainability science, legal studies and political science.

This book is intended for an audience interested in anthropology, political science, Arctic urban studies, youth studies, Arctic social sciences and humanities in general. It would attract those working on Arctic sustainability, wellbeing in the Arctic, Arctic demography and overall wellbeing of youth.

Florian Stammler is Research Professor at the University of Lapland's Arctic Centre (Finland), and coordinates its anthropology team. He led the WOLLIE project consortium "Live, Work or Leave? Youth—wellbeing and the viability of (post) extractive Arctic industrial cities in Finland and Russia".

Reetta Toivanen is a Professor of Sustainability Science at the Helsinki Institute of Sustainability Science (HELSUS), Docent in Social and Cultural Anthropology and the Vice-Director of the Centre of Excellence in Law, Identity and the European Narratives. She is based at the University of Helsinki, Finland.

Routledge Research in Polar Regions

Series Editor: Timothy Heleniak

Nordregio International Research Centre, Sweden

The Routledge series Research in Polar Regions seeks to include research and policy debates about trends and events taking place in two important world regions: the Arctic and Antarctic. Previously neglected periphery regions, with climate change, resource development and shifting geopolitics, these regions are becoming increasingly crucial to happenings outside these regions. At the same time, the economies, societies and natural environments of the Arctic are undergoing rapid change. This series seeks to draw upon fieldwork, satellite observations, archival studies and other research methods which inform about crucial developments in the Polar regions. It is interdisciplinary, drawing on the work from the social sciences and humanities, bringing together cutting-edge research in the Polar regions with the policy implications.

Arctic Sustainability, Key Methodologies and Knowledge Domains
A Synthesis of Knowledge I
Edited by Jessica K. Graybill and Andrey N. Petrov

Food Security in the High North
Contemporary Challenges Across the Circumpolar Region
Edited by Kamrul Hossain, Lena Maria Nilsson, and Thora Martina Herrmann

Collaborative Research Methods in the Arctic
Experiences from Greenland
Edited by Anne Merrild Hansen and Carina Ren

Greenland's Economy and Labour Markets
Edited by Laust Høgedahl

Renewable Economies in the Arctic
Edited by David C. Natcher and Timo Koivurova

Indigenous Peoples, Natural Resources and Governance: Agencies and Interactions
Edited by Monica Tennberg, Else Grete Broderstad, and Hans-Kristian Hernes

Young People, Wellbeing and Placemaking in the Arctic
Edited by Florian Stammler and Reetta Toivanen

For more information about this series, please visit: www.routledge.com/Routledge-Research-in-Polar-Regions/book-series/RRPS

Young People, Wellbeing and Placemaking in the Arctic

Edited by Florian Stammler and
Reetta Toivanen

LONDON AND NEW YORK

First published 2022
by Routledge
2 Park Square, Milton Park, Abingdon, Oxon OX14 4RN

and by Routledge
605 Third Avenue, New York, NY 10158

Routledge is an imprint of the Taylor & Francis Group, an informa business

© 2022 selection and editorial matter, Florian Stammler and Reetta Toivanen; individual chapters, the contributors

The right of Florian Stammler and Reetta Toivanen to be identified as the authors of the editorial material, and of the authors for their individual chapters, has been asserted in accordance with sections 77 and 78 of the Copyright, Designs and Patents Act 1988.

Trademark notice: Product or corporate names may be trademarks or registered trademarks, and are used only for identification and explanation without intent to infringe.

British Library Cataloguing-in-Publication Data
A catalogue record for this book is available from the British Library

Library of Congress Cataloging-in-Publication Data
A catalog record has been requested for this book

ISBN: 978-0-367-62629-7 (hbk)
ISBN: 978-0-367-62630-3 (pbk)
ISBN: 978-1-003-11001-9 (ebk)

DOI: 10.4324/9781003110019

Typeset in Times New Roman
by SPi Technologies India Pvt Ltd (Straive)

Contents

Figures

Tables

Notes on the contributors

Ria-Maria Adams is a PhD candidate at the Department of Social and Cultural Anthropology at the Faculty of Social Sciences, University of Vienna. Her research interests revolve around Arctic youth wellbeing, industrial northern towns and sustainable communities. Since 2018, she has conducted ethnographic research in Rovaniemi, Kolari, Kemijärvi and Pyhäjoki.

Lukas Allemann, PhD, has a background in social anthropology and oral history. In recent years, he researched Indigenous and youth issues in Russia's Northwest. He is based at the anthropology research team at the Arctic Centre, University of Lapland, Rovaniemi, Finland.

Alla Bolotova, PhD, works as a post-doctoral researcher at Aalto University, Finland. She specializes in the fields of Arctic (urban) anthropology and environmental sociology, focusing particularly on youth wellbeing, sense of place and human–environment interactions in industrial communities in the Russian Arctic.

Anna Fomina holds a MA in Sociology from the European University in St. Petersburg. She is a PhD student at the University of Helsinki (Doctoral Programme in Social Sciences). She participates in international projects on child welfare and agencies of young care leavers in Russia.

Susanna Gartler is a PhD candidate based at the Department of Social and Cultural Anthropology of the University of Vienna, and co-investigator with the projects LACE – Labour Mobility and Community Participation in the Extractive Industries – Yukon and Arctic Youth and Sustainable Futures.

Aytalina Ivanova is a Legal anthropologist at the Arctic Centre, University of Lapland and at the Faculty of Law, North-Eastern Federal University, Yakutsk, Russia. Coordinator of the Russian part of the WOLLIE project consortium "Live, Work or Leave? Youth – wellbeing and the viability of (post) extractive Arctic industrial cities in Finland and Russia."

Tanja Joona is Senior Researcher at the Arctic Centre of the University of Lapland, Docent in public international law and at the moment the Finnish Institutional Leader of the H2020 project JustNorth (2020–2023): Toward Just, Ethical and Sustainable Arctic Economies, Environments and Societies.

Pigga Keskitalo is a University Researcher at the University of Lapland, Faculty of Education, Docent in Education at the University of Helsinki, and affiliated researcher at the Sámi University of Applied Sciences.

Teresa Komu holds a PhD in Cultural Anthropology from the University of Oulu. She is currently a Postdoctoral Researcher at the University of Lapland's Arctic Centre, Finland. Her main research interests include land-use conflicts, wellbeing and environmental change in northern Fennoscandia.

Meri Kulmala holds a PhD in Sociology and is Docent in Russian and Eurasian Studies. She works as the director of the interdisciplinary research profiling action Helsinki Inequality Initiative (INEQ) at the University of Helsinki. She has led several research projects on child welfare in contemporary Russia.

Miia Lähde (MSocSc) is a doctoral researcher in social psychology at the Faculty of Social Sciences, Tampere University, Finland. She has worked on several projects in the field of childhood and youth research and as a coordinator of the Finnish University Network for Youth Studies (YUNET).

Taiya Melancon is a First Nation of Nacho Nyäk Dun citizen and resident of Mayo. She is studying in the field of child care, and working with young children and families in hopes to strengthen the foundation of her community.

Jenni Mölkänen holds a PhD in Social and Cultural Anthropology in the University of Helsinki and is a researcher on the ALL-YOUTH research project. She has done long-term fieldwork in Madagascar concerning environmental conservation efforts among subsistence farmers and interviewed young people in Finland about their life course transitions heading towards adulthood and independence.

Tatyana Oglezneva, is head of department of civil law and process North Eastern Federal University Yakutsk, candidate of historical sciences, Associate Professor. She is a member of the WOLLIE project consortium "Live, Work or Leave? Youth – wellbeing and the viability of (post) extractive Arctic industrial cities in Finland and Russia". She worked on the block on youth legislation of the Arctic regions of Russia.

Eileen Peter is a First Nation of Nacho Nyäk Dun citizen, and former member of the heritage department. She now resides in Calgary, where she pursues her passion in graphic design.

Maria Pitukhina is a Doctor of Political Science, Professor at Petrozavodsk State University (Russia) and Research Leader at the Institute of Economy at the Russian Science Academy.

Anna Simakova is a researcher (sociologist, PhD student) from Petrozavodsk State University. She is a member of the WOLLIE project consortium "Live, Work or Leave? Youth – wellbeing and the viability of (post) extractive Arctic industrial cities in Finland and Russia". She was responsible for polling youth in industrial cities in Russia.

Florian Stammler is Research Professor at the University of Lapland's Arctic Centre (Finland), and coordinates its anthropology team. He coordinated the WOLLIE project consortium "Live, Work or Leave? Youth – wellbeing and the viability of (post) extractive Arctic industrial cities in Finland and Russia".

Reetta Toivanen (PhD) is a Professor of Sustainability Science at the Helsinki Institute of Sustainability Science (HELSUS), Docent in Social and Cultural Anthropology and the Vice-Director of the Centre of Excellence in Law, Identity and the European Narratives. She is based at the University of Helsinki, Finland.

Veli-Pekka Tynkkynen is Professor at the University of Helsinki's Aleksanteri Institute (Finland), and coordinates its research group on the Russian environment. He led a subproject within the WOLLIE project consortium "Live, Work or Leave? Youth – wellbeing and the viability of (post) extractive Arctic industrial cities in Finland and Russia".

Acknowledgements

This volume is the combined scholarly outcome of two research projects on Arctic youth carried out in 2018–2020. Most of the contributions are authored by the respective project researchers and the research has been funded by these projects, unless otherwise mentioned in the chapters. The consortium WOLLIE Youth ("Live, Work or Leave? Youth – wellbeing and the viability of [post] extractive Arctic industrial cities in Finland and Russia") was led by Florian Stammler and co-financed by the Academy of Finland (decision numbers 314471 and 314472; PI Veli-Pekka Tynkkynen) and the Russian Fund for Basic Research (project number 18-59-11001; PI Aytalina Ivanova). The consortium ALL-YOUTH (All Youth Want to Rule their World) was led by Reetta Toivanen and financed by the Academy of Finland's Strategic Research Council (decision number 312689). The research has also profited from the Centre of Excellence in Law, Identity and the European Narratives (Academy of Finland decision number 312431), and during the pandemic from the Russian Science Foundation project "Anthropology of Extractivism" (decision number 20-68-46043).

As editors, we thank the authors for their enthusiastic participation in compiling this volume. Even without our planned authors' meeting, which was cancelled due to COVID-19 restrictions, the process was outstanding: in addition to their own writing, authors also reviewed and commented each other's chapters, thus contributing significantly to the volume as a whole. We would also like to thank the additional external reviewers for their valuable comments on the chapters.

Special thanks go to Christine van der Horst for her outstanding assistance throughout the editing process for this volume. Finally, we thank Richard Foley for his extraordinary dedication to our work during his English-language editing of five of the chapters, and to Lukas Allemann for his help with the difficult task of harmonizing bibliographic information and Russian sources and spelling throughout the volume.

Notes on transliteration

In this volume we have used the simplified Library of Congress transliteration system for Russian. For some geographical terms, however, we have opted for transliteration common in the English literature, e.g., Yakutia instead of Iakutiia and Yamal instead of Iamal.

The names of Russian geographical and administrative entities occurring in the texts have been translated as follows:

sub'ekt Federatsii – constituent entity of the Federation
respublika – republic
krai – territory
oblast – region
okrug – district
avtonomnyj okrug – autonomous district
raion – municipality (most cities are also municipalities)

These terms are capitalized where they are part of geographic names, such as Murmansk Region for Murmanskaia Oblast.

Figure 0.1 (Photo: F. Stammler, 2020)

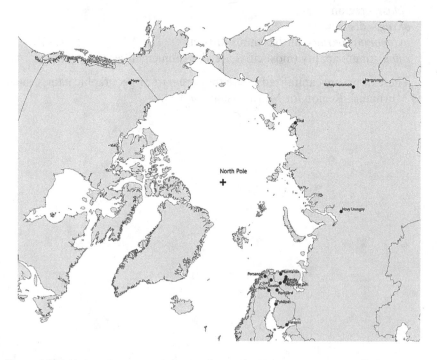

Figure 0.2 Field sites where research for the chapters in this volume was carried out. (Created by: Arto Vitikka)

Introduction

The quest for a good life: Contributions from the Arctic towards a theory of wellbeing

Florian Stammler and Reetta Toivanen

Integrating different theoretical approaches to wellbeing

Anthropological research on wellbeing as a concept has highlighted that studies claiming to be about wellbeing largely fail to identify what it actually means (Thin 2008, p. 36). Ethnographic research from around the world has since started filling this gap (Mathews and Izquierdo (eds) 2008). Jiménez (2008, p. 3) emphasizes the importance of ethnography for informing our understanding of wellbeing, especially now that the concept has gained more political attention in a globalizing world. However, so far there is a lack of research focusing on wellbeing in the Arctic, and this is where this volume seeks to close a research gap.

Ortner (2016) suggested in a seminal article on dark anthropology that in order to understand the foundations of a good life, we need to study more than the absence of harm and hardship. In this volume, we want to provide empirical evidence on how good life in the Arctic looks from the point of view of its young residents. In that vein, this introduction is an attempt to conceptualize such evidence in theoretical terms. As Robbins (2013) has shown, an anthropology of the good life provides positive definitions of the parameters 'of what is good' but without overlooking the dark side of life. This should contribute to clarifying more generally how we can better define wellbeing in the social sciences.

In this respect, the anthropology of a good life aligns with recent trends in human security studies, for which wellbeing became a concept that invited scholars to define what is called "positive security" (Hoogensen et al. 2009; Hoogensen Gjørv 2012). In that field, a good life is first and foremost associated with the absence of harm and threat. Even in critical security studies where the focus is on security *to* (to satisfy one's needs, to live one's routine) and not security *from* (from threats, violence or harm), researchers assume that a good life is one that enables people to cope with risks and danger and satisfies their needs against all odds (see overview in Hoogensen Gjørv 2012, pp. 835–843; Stammler et al. 2020). What is called positive security in human security studies (Hoogensen Gjørv 2012, pp. 835, 843–846) comes closest to what anthropologists have highlighted as relevant in the study of wellbeing (Jiménez 2008).

DOI: 10.4324/9781003110019-1

Both fields (anthropology of wellbeing and positive human security studies) point to the importance of considering contexts, values, justice, equality and trust for the understanding of wellbeing and human security. At the same time, they both highlight the need to comprehend practices and processes of achieving these qualities among people. Also, both anthropologists and human security scholars highlight the difficulties of defining these positive qualities. One does not need to be a profound philosopher to realize that it is harder to define what is good without knowing what is bad, or to outline a clear research programme without the background of a gap analysis. This comprises the rather simple underpinning of the arguments by Robbins (2013) and Thin (2008) when they start outlining what studies of the good life are missing. Hoogensen Gjørv (2012, p. 843) takes an "epistemology of enabling" as a starting point to define what good may mean in her field of positive security studies: "Through positive security, people in specific contexts are recognized as potentially having some significant resources to tackle challenges and risks" (Hoogensen Gjørv 2012, p. 844). From this quote, the reader again gets the impression that the focus is on coping strategies, enabling the overcoming of problems, harm, risk and threat, and achieving freedom from these negative things through trust, justice and equality. This approach does not yet tell us what the positive parameters of wellbeing actually are, other than the capacity to overcome problems.

Equality and justice are fundamental for Jiménez (2008, p. 4) in the introduction to his volume that studies the place of wellbeing in contemporary theories of political morality. Jiménez reminds us of Sen's influential idea of wellbeing as an attempt for a more affirmative definition. It goes beyond the satisfaction of basic needs and rights to encompass substantive freedoms and basic capabilities, and incorporates both the opportunities that people have to change their lives and the processes that they undertake to harness such opportunities (Sen 1999, quoted in Jiménez 2008, p. 8). However, this still leaves us with a number of questions: What actually is it that people should have the freedom to achieve? What are the dreams that people should have the opportunity to pursue? The research in this volume shows that there is more to wellbeing than variables of the Human Development Index, which Sen helped to develop, and that especially young people in remote places such as the Arctic may not feel represented by such a broad and all-encompassing notion of wellbeing as Sen's (1999, quoted in Jiménez 2008, p. 8), which focuses on political freedom, economic facilities, social opportunities, transparency guarantees and protective securities. Thin's (2008) scholarly quest that wellbeing studies should define the positive parameters of a good life thus remains largely unanswered by the earlier literature.

In their volume on youth wellbeing from the point of view of the educational sciences, Wright and McLeod (2015) observed that in the attempt to define wellbeing, international rankings usually focus on what they call "objective measures" such as GDP, expenditures for health, education and the like. On the other hand, they state that national policies see wellbeing more in terms of social and emotional dimensions, particularly psychological notions and mental health (p. 2). The chapters in this volume do not fall so easily into the simplistic separation of "objective" and "subjective" indicators of wellbeing that Wright and McLeod (2015, p. 2) categorized. Our approach to youth wellbeing is that thick ethnographic descriptions can best show what is important for young people's good life in the Arctic, as well as what the factors are that may influence their decisions to pursue their dreams and opportunities in the Arctic or elsewhere. Wyn et al. (2015) have argued that wellbeing is often defined as an imperative by society. Hence, they argue that the parameters of wellbeing or the scales to measure it are not made by young people and are beyond their control. Through our research we have sought to correct this. We have placed the young people in the centre of our inquiries and worked together with Arctic young people in order to be able to contribute in-depth and multi-year fieldwork-based empirical evidence on how young people in the Arctic see a good life.

A significant body of literature on the concept of wellbeing comes from the area of mental health studies and psychology (Ryan and Deci 2001; Usborne and Taylor 2010), where the focus is on the individual rather than the group as determining people's sense of a good life. Since the present volume places the focus more on groups than individuals, we do not attempt to exhaustively engage this literature. Nonetheless, it is important to link it to our discussion, because psychological aspects figure prominently in the literature on Arctic wellbeing, which is strongly oriented towards Indigenous mental health and psychological aspects (Kral et al. 2011; Rasmus et al. 2014; Ulturgasheva et al. 2014; Ulturgasheva et al. 2014; Petrasek et al. 2015; Hatala et al. 2017; see also Gartler et al., this volume).

In psychology, there is a strand of research called "wellbeing science" (Oades et al. 2021). While Oades et al. claim that their approach is highly interdisciplinary, they place crucial importance on the individual alone. They write: "Wellbeing is highly individual and the freedom and choice to decide what wellbeing means to them, is essential to wellbeing itself" (Oades et al. 2021, pp. 719–720). They further develop Sen's (1999, cited in Jiménez 2008) idea of capability and subjective freedom, emphasizing that the individual's capacity to choose their own parameters of wellbeing is crucial. For the authors, there are critical universal building blocks for the capability to thrive, based on what a person needs to be free *from* and free *to do* for a lived experience of wellbeing. Accordingly, individual wellbeing is based on freedom *from* poverty and instability, disease, alienation and isolation, violence and corruption, and freedom *to* choose one's own life trajectory. According to Oades, that freedom *to* is less universal than the freedom *from*, which is why we can uncover the diversity of that freedom *to*, the positive definition of wellbeing best understood by analysing different narratives and what he calls "differentiated life-stories" (Oades 2018, see video at 36:58). Building on these ideas, our contribution in this volume is an in-depth study of some of the factors that influence such life trajectories, not only on the individual level but also on the group level, necessitated by structural factors such as age (hence our focus on youth) and regional specifics (here the Arctic). This is especially relevant, as neither anthropology nor geography inform Oades' interdisciplinary approach to wellbeing, which he calls thriveability theory (Oades 2018, see video at 11:30; Oades et al. 2021).

Wellbeing as an applied category for measuring quality of life

Different from the theoretical political, philosophical and psychological literature discussed above, a sizeable part of wellbeing studies is applied and appears as white papers or reports targeted at policymakers. For the Arctic, such studies include, for example, Lundgren and Cuadrado (2020) on skills in the North. Different from the fieldwork-based guide recently published by some of the authors of this volume (Adams et al. 2020), Lundgren and Cuadrado (2020) take well-known wellbeing indices as their point of departure and analyse the good life in the European North using statistics of life expectancy and socio-economic factors, of which they single out education, gender equality and mental health. They also cite a high level of social trust as characteristic of a specifically northern European parameter of wellbeing, which echoes Hoogensen Gjørv's (2012) emphasis on trust for positive human security. Quite differently, the study by Ingemann and Larsen (2018) has a strong focus on what we could call "deficit analysis", analysing the literature to establish where and why the Arctic region loses out in comparison to other places in terms of conditions for providing a good life for young people. Hence, in such reports the concept of wellbeing serves to inform policies and correct deficits. We start from the other side: our research has demonstrated that young people in the circumpolar North do not necessarily see themselves as deficient, underprivileged, marginalized or disempowered. Rather, there are positive conditions and parameters for their wellbeing that they find in their northern home places.

The chapters in this book show what these conditions and parameters for wellbeing are and how they are vary between various regions in the Arctic. We take wellbeing as an analytical category down to the level of the everyday life of young people in remote places, and investigate how they imagine a good life and what their place of living could offer to them to facilitate achieving their dreams. The wellbeing of some may be influenced by seemingly trivial things, such as having more shopping malls in one's vicinity, or places to hang out with friends in a relaxed and safe atmosphere, or access to a beautiful and clean natural environment, which for some serves as inspiration and recreation in their daily life. Allemann's research (Adams et al., this volume) has summarized some of these aspects as a quest among young people in the Arctic for more hedonistic opportunities rather than eudaemonic wellbeing. We suggest that such parameters of wellbeing be included more in future analysis of what makes young people in the Arctic feel good. Of course, this goes hand in hand with other crucial parameters of wellbeing, which figure as prominently in our research results as in those of international quality of life metrics, such as housing, employment and education, as shown in Table 0. What unites these factors is that they provide a more down-to-earth operationalization of what a good life may mean in everyday life.

Table 0.1 Parameters of wellbeing operationalized for field research with Arctic youth in Eurasia, compared to the parameters of the OECD Better Life Index

Parameters of wellbeing used for guiding fieldwork with youth in the Arctic (2018–2020)	Parameters of wellbeing in the OECD Better Life Index
1. nature, climate and environment; 2. transport and mobility, distance to big centres and convenience of small cities; 3. quality of life due to housing, medicine and a healthy environment; 4. economic perspectives: labour market, career opportunities; 5. social fabric of the community, networks and openness of civil society for youth; 6. locally appropriate education plus opportunities for education in metropolitan centres; 7. availability and diversity of services and opportunities for finding one's own niche as part of society; 8. degree to which legislation corresponds to actual needs and desires expressed by youth; 9. quality and diversity of spending free time (e.g. culture, education, nature, sports); 10. safety/security in the city for young people, including young families; 11. pride in "northernness" or similar local loyalties.	1. housing; 2. income; 3. jobs; 4. community; 5. education; 6. environment; 7. civic engagement; 8. health; 9. life satisfaction; 10. safety; 11. work-life balance.

We believe in the usefulness of this type of more down-to-earth approach, as some very relevant literature on wellbeing remains on abstract levels of political philosophy or complex social theory. What do we really gain from such equivalencies as 'positive' equals 'good' or 'just'(Hoogensen Gjørv 2012, p. 845), defining "wellbeing as simply living well" (Oades et al. 2017, p. 99), or understanding wellbeing as a complex interplay between proportionalities and limits (Jiménez 2008)?

The findings in the chapters of this volume gain more relevance against the backdrop of notions of wellbeing that have arisen out of different applied wellbeing indicators. Among numerous such efforts, we found the OECD's better life index to be particularly interesting. This initiative combines measurable variables and perceptions in an online tool by inviting people to rate the parameters of wellbeing according to their own importance. We fully subscribe to the starting point and first sentence on the OECD homepage (OECD Better Life Index n.d.): "There is more to life than the cold numbers of GDP and economic statistics." Many of the variables for wellbeing that website visitors are invited to assess mirror our own research results, presented in this volume. The OECD tool combines the votes of the participants into country averages of wellbeing on a 1–10 scale, with all the Arctic countries (except Russia) appearing above the OECD average. Rather than confirming or challenging such quantitative results, in this volume we focus on qualitative thick descriptions of how the parameters of wellbeing unfold among our research partners in various places in the Arctic. The parameters that informed a significant amount of the fieldwork conducted for the chapters in this volume at first glance look very similar to those used by the OECD, as Table 0.1 indicates.

The similarity of the two lists indicates that as humans grow up to become adults, there are some universal parameters that will influence their quality of life, regardless of the region, country, political system and culture in which they grow up. While the two lists show that it is not so much the parameters of wellbeing that are specific to the Arctic, we argue that some of them acquire a different meaning in the region, particularly due to the influence of two principal factors: climate/environment and transportation/mobility.

Why the Arctic?

We would like to stress climate and the environment of Arctic settlements as two overarching factors that influence all other parameters of human wellbeing in the region. Thus, they cannot be left outside of considerations on the attractiveness of the North as a place for young people to pursue their plans and dreams in life. Different from more temperate regions, everybody in the Arctic agrees on the strong influence of the environment and climate on quality of life.

In more temperate climates, the difference between the seasons allows for a life independent of the environment. In the Arctic, most people's day starts with taking a look at the temperature, precipitation and daylight, which are more diverse throughout the year than anywhere else on the planet. Long periods of snow and darkness in the winter and 24/7 daylight in the summer can be a source of depression or inspiration, but either way they are influential and impossible to ignore, compared to the case of a temperate metropolitan area. This gives the parameter of the environment a specific meaning in the Arctic.

Moreover, the fact that the Arctic is remote from the various countries' capital and metropolitan areas is a parameter that is constantly mentioned, even more so by young people. Connected to this is the parameter of transportation, the ability to explore the surrounding nature and connect with friends using different vehicles in the Arctic, or experiencing the distance to big centres as detrimental to the sense of quality of life. Arctic settlements are small and far away from the hustle and bustle of busy life. Some consider this an asset, enabling calm and peace of mind, while for others it is experienced as a deficit of opportunities.

These two overarching factors decisively influence the rest of the parameters on the list. For example, in smaller Arctic settlements many mention the density of social networks and the support of family, friends and neighbours as different from the anonymity of life in a big city. That can also be considered a function of spatial distance to populous centres, as well as a harsh environment where support by neighbours used to be a matter of survival, with services being less institutionalized than in large cities. Likewise, the factors mentioned above figure prominently as parameters of local identity, which may support young people's decisions to either pursue their life in the North, return back after education in southern areas (Adams et al., this volume), or serve as a sense of belonging even among those who have left (Toivanen, this volume).

Given the focus mentioned above in wellbeing studies on Indigenous peoples, psychology and mental health in North America (Kral et al. 2011; Rasmus et al. 2014; Ulturgasheva et al. 2014), with our regional focus on the Eurasian Arctic in this volume we complement the existing literature, without leaving out North America (see Gartler et al., this volume). Additionally, we specifically do not categorize our research participants according to ethnic principles.

Why youth?

Youth obviously has a long history as a defining category for social science research, at both the individual level (psychology) and the collective level (sociology and anthropology). The classical study by Mead (1928) and its later reception have been formative for our understanding of the cultural specifics of youth as an age-class, as well as the importance of context in region and culture. This work and its reception have also influenced for a long time the way in which anthropologists have related to youth, mainly as the age category before initiation into adulthood, along with sexual practices and youngster male/female relationships (Bucholtz 2002). Obviously, youth as an age group is more than adolescence.

It does not come as a surprise that the more that studies become detailed and in-depth, the more we can uncover the diversity within this age group. Teenagers have different priorities for their wellbeing than young adults, school pupils, young professionals and just-married couples, to name just a few different phases of youth. The cultural practices of these differ not only from adults but also within youth. Bucholtz (2002, p. 525) observes that, more recently, research on youth has become wider and produces more studies on youth cultural practices. We find this orientation particularly important, because understanding these cultural practices tells a great deal about the future of Arctic societies and settlements. In this volume, we seek to integrate fields of Arctic social sciences that often separate between Indigenous, incomer, nomadic, settled, rural and urban communities. Focusing on youth allows us to bridge this divide.

Wright and McLeod (2015, p. 4) have highlighted that youth is a volatile category of the human population. Understanding variation and influences on this volatility is even more important, as it is the young generation that shapes the future and viability of human presence in the Arctic at a time when the region is going through substantial changes, such as global warming and extractive industrial development. From the perspective of both the public and natural science, this has made the Arctic into a natural laboratory where many planetary developments can be observed at a faster pace and are more clearly expressed than in more temperate regions. With this volume, we argue that if the future of the Arctic is of interest for the whole world, we must understand the motivations, agency and cultural practices of the generation of inhabitants that will shape the future of that region from within.

Mobility, agency and regulating paths to independence: a road map for the volume

We will now outline the main findings of the chapters in this volume and highlight several overarching topics that run through them. The volume has benefitted from all the writers being involved in reading, commenting and cross-referencing each other's contributions. During this process we observed that mobility and emplacement, youth agency, regulative practices of youth life and paths to independence figure particularly prominently in many chapters. Correspondingly, this volume is divided into sections with these headings.

Mobility and emplacement concern both physical and social mobility, within the Arctic and between the Arctic and more southern regions. Many Arctic settlements, particularly in Russia but also in Finland, are losing their populations. This is the background situation on which the chapters by Simakova et al., Bolotova, Komu and Adams, and to some extent also Oglezneva et al. and Toivanen, develop their discussions. Though having different theoretical interests and disciplinary orientations, all of them find that for young people in Finland's and Russia's Arctic, the default situation is one where a young person moves away from the North after leaving the parents' household. In Chapter 1, Simakova, Pitukhina and Ivanova highlight the importance of understanding youth's idea of wellbeing in relation to their migration intentions out of Arctic single-industry towns in Russia. They distinguish between the economic, social and emotional building blocks of wellbeing, based on a sociological survey that they conducted. With this trifold division, they combine what Wright and McLeod (2015, p. 2) have observed as being separate, namely, that many wellbeing metrics rely on numerical values, while wellbeing policies are based on psychological and emotional variables. Simakova et al. conclude that "migration sentiments" influence young people's decisions to leave or stay in Arctic single-industry towns.

On the Finnish side, in Chapter 2 Komu and Adams rightly remind that youth outmigration is part of a trend towards urbanization which is not only characteristic for the Arctic but for rural areas in general. They highlight, in particular, that moving out is the default situation for young people, while staying behind is less prestigious and feels like "being stuck". Therefore, based on their case study of the small post-industrial town of Kolari in Finnish Lapland, they develop their main idea of a "culture of migration", where mobility in everyday life and the freedom to move out are important for young people's notion of wellbeing. In Chapter 3, Bolotova speaks to exactly the same ideas, but based on long-term fieldwork in single-industry towns in Russian Lapland. She observes that there is much less research on "stayers" than on "away-movers". Her work with stayers shows that there is more to remaining in Arctic towns than what she calls "involuntary immobility" and that staying is not a one-time act but a process with diverse facets and nuances. In her ethnography of staying, Bolotova therefore highlights the agency not only of migrants but also those young people who choose to stay in the Arctic, thus contributing to a more positive notion of a good life.

(Im)mobility and agency are also the two key terms around which Toivanen builds her argument in Chapter 4. Rather than focusing on stayers, however, she identifies a strong sense of belonging in the North among those Indigenous youth that moved out of the Arctic to more southerly areas, such as the Finnish capital of Helsinki. This shows, in particular, young people's agency to stay connected to culturally significant places and livelihoods even in cases when they are physically distant from them as a result of outmigration.

In Chapter 5, Joona and Keskitalo focus on those young people who practice a specific livelihood that continues to exist in the Finnish and Swedish Arctic: reindeer herding. Their findings allow the question of rural outmigration and involuntary immobility to appear in a different light than the previous chapters, portraying on the one hand a bleak picture of villages and towns emptying out, a lack of employment options, the low qualification of the stayers, marginalization, dissatisfaction, violence and abuse. On the other hand, they emphasize their young interlocutors' strong connection to the Arctic environment through reindeer herding and their agency and decisiveness to continue their livelihood. The authors can also be especially lauded for highlighting the gender dimensions of such choices: boys are more likely to take on reindeer herding as their primary profession. Joona and Keskitalo end with the hopeful finding that girls are increasingly considering reindeer herding, at least as a part-time profession.

Many reindeer herders consider themselves Indigenous, and Indigenous youth is also the topic of Chapter 6, in which Gartler, together with Melacon and Peter, focus on the Yukon in Canada. While reindeer herding as a livelihood is not an option there, what is called "living off the land" clearly emerges as a source of wellbeing for Indigenous youth, even though the region is nowadays dominated not only by extractive industries but also extractivism as an approach to life. Gartler shows that income from mining enhances the possibilities of Indigenous youth to spend time on the land by enabling the purchase of expensive equipment, which is needed nowadays to access remote places in the Arctic. This is particularly important, as being on the land for these young people also means maintaining relations with like-minded human and non-human persons in the environment. Thus, Gartler argues, the impacts of extractive industries for Indigenous youth cannot be seen as solely negative or positive.

As Wyn et al. (2015) have argued, the parameters of wellbeing as imperative by society are not made by youth themselves. This is best exemplified in regulative settings and youth policy, in which young people's participation is minor. Such rules and laws often have direct effects on the key condition of youth wellbeing, as identified by many authors (Jiménez 2008; Oades et al. 2021).

In Chapter 7, on Russian youth law and politics in the Arctic, Oglezneva, Ivanova and Stammler give examples of youth agency in the determination of their wellbeing in the Arctic (for example, through youth parliaments, youth policy programmes and civil society initiatives). Their findings show the diversity of approaches and situations in the largest Arctic country, which is also due to the absence of a Russian federal law on youth and on the Arctic. Using research evidence from fieldwork in three northern industry towns, the authors reveal how certain regions and municipalities have a significant ability to implement their own policy in order to make themselves attractive to youth, even though for outside observers Russia may seem as a centrally administered country where decisions are mainly made in Moscow. They show that this local and regional power is also strongly facilitated by the youth social policies of big industrial companies that are key economic actors in Russian Arctic cities. This argument reminds us that governance has long moved beyond the state being the only actor, and youth policy in the Arctic is in line with the recent trend towards multi-actor, multi-level governance, for which the Arctic Council has become famous (Hoogensen Gjørv 2012, p. 865).

After all, youth policy should enable young people's paths to independence and the shaping of their own future, as well as that of the region. Those young people who have lived in institutional care have special challenges in seeking independent adulthood. Chapter 8 by Lähde and Mölkänen explores narratives of the independence of young adults who have been clients of youth welfare services in Finland. They discuss three predominant common themes (insecurities in social relations, illness or struggles with psychosocial wellbeing, and moving) that manifested in young adults' narratives, and consider how these contribute to the needs and possible spaces for support based on young adults' experiences. With a similar research orientation but set in Russian northern alternative care, in Chapter 9 Kulmala and Fomina explore the expectations of young people who transition from different forms of alternative care into their independent adult life. Their empirical analysis is structured by two modes of future orientation by the young adults in the study: those who plan and dream ahead and those who show little future orientation or a refusal to plan.

In Chapter 10, the final chapter of this volume, Adams, Allemann and Tynkkynen make a Finnish–Russian northern comparison: their two case towns, Pyhäjoki and Polyarnye Zory, are united by a crucial corporate agent, the Russian state's nuclear company Rosatom, which runs one plant in Russian Lapland and has partnered with a Finnish company to build another one in Pyhäjoki in northern Finland. The authors highlight the importance of corporate agency for youth wellbeing, particularly in relation to evidence from a single-industry town in Russia, with influence still being much less on the Finnish side.

Conclusion

This introductory chapter has highlighted the potential of in-depth, empirically grounded research with young people in the Arctic to contribute to theoretical and applied notions of wellbeing in the social sciences. We have shown some ways in which literature from such different disciplines as anthropology, psychology, human security studies and educational science can become relevant for an integrated understanding of youth wellbeing, using evidence from the Arctic. Building on the dominant theoretical ideas of *capability* and the *freedom to act* for one's future, this introduction and the chapters of this volume flesh out what the crucial empirical parameters for a good life in the Arctic are for young people—in other words, what Arctic youth are capable of doing to achieve the kind of wellbeing that would make them like their life in connection to their homeland.

We have shown that some of these parameters seem similar to those in the dominant global wellbeing indicators, such as the OECD Better Life Index. However, the crucial factors of the Arctic climate/environment, the specifics of its geography and the connected questions of mobility cause these parameters to play out in ways that make the Arctic different from other places on our planet. Highlighting the agency of youth with examples from the Arctic, this volume shows that it is not enough for youth to wait until their seniors create more favourable conditions.

Arctic youth must be confident in their economic and social potential and take an active position, creating their own future and that of their children. The chapters in this book give a strong positive signal that young people have all the capacity and abilities needed to actively take part in shaping their own lives. There are manifold governmental and economically motivated plans to 'develop' the Arctic, which render the Arctic inhabitants invisible and irrelevant (see Toivanen 2019), but the empirical and ethnographic chapters in this book significantly challenge these century-long narratives of the Arctic as a peripheral resource frontier. Arctic youth do not need to make the same choices as earlier generations, because due to technology the places of work and education and the places of family can be connected. Thus, the Arctic has a lot to offer to young people searching for a good life, if they wish to locate themselves in the region.

References

Adams, R.-M., Allemann, L. and Tynkkynen, V.-P. (this volume) 'Youth well-being in "atomic towns": The cases of Polyarnye Zori and Pyhäjoki', in Stammler, F. and Toivanen, R. (eds) *Young people, wellbeing and placemaking in the Arctic*. London: Routledge, pp. 222–240.

Adams, R.-M. et al. (2020) *'Ensuring thriving northern towns for young people: A best practices guide / Vetovoimainen kunta pohjoisen nuorille – parhaita käytäntöjä / Molodezh' v severnykh gorodakh: luchshie praktiki'*. Arctic Centre, University of Lapland. Available at: https://www.arcticcentre.org/EN/Youth-and-Wellbeing.

Bucholtz, M. (2002) 'Youth and cultural practice', *Annual Review of Anthropology*, *31*(1), pp. 525–552. doi: 10.1146/annurev.anthro.31.040402.085443.

Gartler, S with Melanon, T. and Peter, E. (this volume). 'Indigenous youth perspectives on extractivism and living in a good way in the Yukon', in Stammler, F. and Toivanen, R. (eds) *Young people, wellbeing and placemaking in the Arctic*. London: Routledge, pp. 120–146.

Hoogensen Gjørv G. et al. (2009) 'Human security in the Arctic-yes, it is relevant!', *Journal of Human Security*, *5*(2), pp. 1–10. doi: 10.3316/JHS0502001.

Hoogensen Gjørv, G. (2012) 'Security by any other name: Negative security, positive security, and a multi-actor security approach', *Review of International Studies*, *38*(4), pp. 835–859. doi: 10.1017/S0260210511000751.

Ingemann, C. and Larsen, C. V. L. (2018) *A scoping review: Well-being among Indigenous children and youth in the Arctic – with a focus on Sami and Greenland Inuit*. Copenhagen: Nordic Council of Ministers. Available at: http://urn.kb.se/reso lve?urn=urn:nbn:se:norden:org:diva-5151 (Accessed: February 17 2021).

Jiménez, A. C. (2008) 'Introduction: Wellbeing's re-proportioning of social thought', in Jiménez, Alberto Corsin (ed.) *Culture and well-being: Anthropological approaches to freedom and political ethics*. London: Pluto Press, pp. 1–32. doi: 10.2307/j. ctt18fs330.4.

Kral, M. J. et al. (2011) 'Unikkaartuit: meanings of well-being, unhappiness, health, and community change among Inuit in Nunavut, Canada', *American Journal of Community Psychology*, *48*(3–4), pp. 426–438. doi: https://doi.org/10.1007/s10464-011-9431-4.

Lundgren, A. and Cuadrado, A. (2020) 'Wellbeing in the Nordic region', in Ludgren, A., Randall, L. and Norlén, G. (eds.) *State of the Nordic region 2020*. Copenhagen: Nordic Council of Ministers, pp. 130–140. Available at: http://urn.kb.se/resolve?ur n=urn:nbn:se:norden:org:diva-6213 (Accessed: July 29 2021).

Mathews, G. and Izquierdo, C., eds (2008) *Pursuits of Happiness: Well-Being in Anthropological Perspective*. Oxford, New York: Berghahn Books.

Mead, M. (1928) *Coming of age in Samoa: A psychological study of primitive youth for Western civilisation*. New York (NY): HarperCollins.

Oades, L. G., Deane, F. P. and Crowe, T. P. (2017) 'Collaborative recovery model: From mental health recovery to wellbeing', in Slade, M., Oades, L., and Jarden, A. (eds) *Wellbeing, recovery and mental health*. Cambridge: Cambridge University Press, pp. 99–110. doi: 10.1017/9781316339275.010.

Oades, L. G. (2018, 22 November) *MGSE Deans Lecture – How do we learn to thrive? The emergence of wellbeing science*. Available at: https://www.youtube.com/ watch?v=ocqFzGjMDa4 (Accessed: February 17 2021).

Oades, L. G. et al. (2021) 'Wellbeing literacy: A capability model for wellbeing science and practice', *International Journal of Environmental Research and Public Health*, *18*(2), pp. 719–731. doi: 10.3390/ijerph18020719.

OECD Better Life Index (n.d.) . Available at: http://www.oecdbetterlifeindex.org/ (Accessed: February 17 2021).

Ortner, S. B. (2016) 'Dark anthropology and its others: Theory since the eighties', *HAU: Journal of Ethnographic Theory*, *6*(1), pp. 47–73. doi: 10.14318/hau6.1.004.

Petrasek MacDonald, J., Cunsolo-Willox, A., Ford J., Shiwak, I., Wood, M. (2015) 'Protective factors for mental health and well-being in a changing climate: Perspectives from Inuit youth in Nunatsiavut, Labrador', *Social Science & Medicine*, *141*, pp. 133–141. doi: 10.1016/j.socscimed.2015.07.017.

Hatala, A. R. Pearl, T., Bird-Naytowhow, K., Judge, A., Sjoblom, E., Liebenberg, L. (2017) '"I Have Strong Hopes for the Future": Time Orientations and Resilience

Among Canadian Indigenous Youth', *Qualitative Health Research, 27*(9), pp. 1330–1344. doi: 10.1177/1049732317712489.

Rasmus, S. M., Allen, J. and Ford, T. (2014) "Where I have to learn the ways how to live:" Youth resilience in a Yup'ik village in Alaska, *Transcultural Psychiatry, 51*(5), pp. 713–734. doi: 10.1177/1363461514532512.

Robbins, J. (2013) 'Beyond the suffering subject: Toward an anthropology of the good', *Journal of the Royal Anthropological Institute, 19*(3), pp. 447–462.

Ryan, R. M. and Deci, E. L. (2001) 'On happiness and human potentials: A review of research on hedonic and eudaimonic well-being', *Annual Review of Psychology, 52*(1), pp. 141–166. doi: 10.1146/annurev.psych.52.1.141.

Stammler, F., Hodgson, K. K. and Ivanova, A. (2020) 'Human security, extractive industries, and Indigenous communities in the Russian North', in Hoogensen Gjørv, G., Lanteigne, M. and Horatio, S.-A. (eds) *Routledge handbook of Arctic security*. London: Routledge, pp. 377–391.

Thin, N. (2008) 'Why anthropology can ill afford to ignore well-being', in Mathews, G. and Izquierdo, C. (eds) *Pursuits of happiness: Well-being in anthropological perspective*. Oxford, New York: Berghahn Books, pp. 23–44.

Toivanen, R. (2019) 'European fantasy of the Arctic region and the rise of Indigenous Sámi voices in the global arena: integrating leadership, discernment and spirituality', in Kok, J. and Van den Heuvel, S. (eds) *Leading in a VUCA world integrating leadership, discernment and spirituality*.Cham, Switzerland: Springer International Publishing, pp. 23–40. doi: 10.1007/978-3-030-05523-3_3.

Toivanen, R. (this volume) 'Towards a sustainable future of the Indigenous youth: Arctic negotiations on (im)mobility', in Stammler, F. and Toivanen, R. (eds) *Young people, wellbeing and placemaking in the Arctic*. London: Routledge, pp. 77–92.

Ulturgasheva, O. et al. (2014) 'Arctic Indigenous youth resilience and vulnerability: Comparative analysis of adolescent experiences across five circumpolar communities', *Transcultural Psychiatry, 51*(5), pp. 735–756. doi: 10.1177/1363461514547120.

Usborne, E. and Taylor, D. M. (2010) 'The role of cultural identity clarity for self-concept clarity, self-esteem, and subjective well-being', *Personality and Social Psychology Bulletin, 36*(7), pp. 883–897. doi: 10.1177/0146167210372215.

Wright, K. and McLeod, J. (eds) (2015) *Rethinking youth wellbeing: critical perspectives*. Singapore: Springer. doi: 10.1007/978-981-287-188-6.

Wyn, J. et al. (2015) 'The limits of wellbeing', in *Re-thinking youth wellbeing: critical perspectives*. Singapore: Springer, pp. 55–70.

Part I

Movement and emplacement

1 Motives for migrating among youth in Russian Arctic industrial cities

Anna Simakova, Maria Pitukhina and Aytalina Ivanova

Introduction: youth in the urban Russian Arctic

The Russian Arctic is strategically important for the country's development, as most of the natural resources that fuel the economy are located there. However, despite its strategic status, the Russian Arctic is characterized by a sharp asymmetry in social and economic development: the region encompasses some of the country's economically most advanced territories as well as some of its most depressed.

The share of youth is lower in the Russian Arctic (13.4%) than the Russian average (19%). The Russian government has recently agreed on an ambitious strategy with development goals for the Russian Arctic (Russian Federation, President 2020). According to the document, the achievement of the strategic goals will require both the retention of young people in the Arctic and attracting others from elsewhere in the country. The demographic indicators in Arctic regions show a steady downward trend in the population for the period until 2035 due to a decrease in fertility, an increase in childbearing age and a decrease in the number of marriages. This situation is further aggravated by youth outmigration. According to the strategy, 200,000 new jobs will be created in the Russian Arctic by 2035, particularly in the Yamal-Nenets Autonomous District, Krasnoyarsk Territory, and the Murmansk and Arkhangelsk Regions (Russian Federation, President 2020). The youth in the Arctic represent an important potential source of labour in implementing the planned large-scale investment projects.

The population of the Russian Arctic zone is largely urban: 89 per cent of the population live in cities, but the proportion of youth in the population is 6 per cent lower than the national average (Statistics Russia 2018). In this chapter we discuss the problems arising from a lack of young people in Arctic industrial cities. Most of the cities are of the single-industry type ("monocities"), meaning that they were built during the Soviet Union to host the workforce needed to extract or process the natural resources of the Arctic. Today, a significant proportion of the cities have the status of "being in the most difficult social and economic situation" or "facing risks of a worsening of the social and economic situation" (Government of the Russian Federation 2014a). What is more, only 22.3 per cent of all Russian

DOI: 10.4324/9781003110019-3

monocities demonstrate sustainable development, defined in this research as "a shifting process in which resource exploitation, redistribution of investments and technological development go in harmony with social wellbeing and environmental balance, providing added value to both current and future potential" (Statistics Russia 2018).

As noted by Mart'ianov (2013, p. 125), in Russia "almost all northern Arctic settlements are artificially created monocities, which makes them vulnerable in the open economy". The single-industry style of urban development makes it difficult to diversify the range of citizens' activities and to increase options for developing labour markets, education and leisure. The variety of leisure activities for young people in such communities is limited compared to that found in large cities and regional centres (Monogoroda Arkticheskoi zony RF 2016). This may be one of the main reasons for the outflow of the population from monocities and, in particular, offer a compelling reason of why young adults leave them. Studies of Arctic monocities reveal that they "predominantly show a regression in demographic indicators: there is a negative population dynamic, the migration outflow is increasing and only a quarter of the cities have recorded a natural population growth" (Statistics Russia 2018).

The statistical indicators characterizing the youth as a social group in the regional communities offer insights into the problem. According to the Russian government, young people in Russia are defined as those aged between 14 and 30 years (Government of the Russian Federation 2014b). Their proportion in terms of the total population of the Russian Arctic is 13.8 per cent on average, compared with 19.3 per cent in the Russian Federation as a whole (Figure 1.1). It is only in the Arctic zone of the Republic of Sakha (Yakutia), Krasnoyarsk Territory and Arkhangelsk Region that as much as one in five residents is a young person. In Murmansk Region and the Yamal-Nenets and Chukotka Autonomous Districts, the

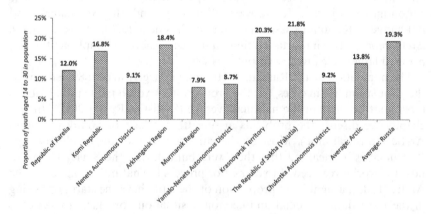

Figure 1.1 Proportion of young people between the ages of 14 and 30 years in the total population of the Russian Arctic regions.

proportion of young people is significantly lower than in the country as a whole. These regions face a declining birth rate as well as an outflow of youth in pursuit of increased educational opportunities.

The Russian Arctic exhibits a number of specific features, such as harsh climatic conditions, their remoteness from federal or regional centres and a high degree of dependence on single branches of industry (primarily the extractive industries). These may contribute to a high likelihood that young people will leave the regions, given their relatively high levels of mobility. The Russian Arctic experiences intensive migration flows, in terms of both arrival and departure of the population from the region (Fauzer 2016). On the other hand, regardless of their location, youth are the most mobile part of the country's population anywhere, ready to pursue ambitious goals in life. Timothy Heleniak claims, in his paper "Migration in the Arctic", that due to the small size of the population and settlements, migration plays a significant impact on the overall population in the Arctic (Heleniak 2014). The following discussion on broad migration flows is a disaggregation of those flows in terms of the key factors of age, gender and level of education.

Ulrich Beck, the pioneer of the concept of the risk society, noted that society is moving toward a new stage of modernity that produces risks. In his words, a risk is defined as a "systematic way of dealing with hazards and insecurities induced and introduced by modernization itself" (Beck 1992, p. 21). According to Beck, some people are more affected than others by the distribution and growth of social risks. Our research draws on the concept of the risk society to investigate the phenomenon of social risk, which occurs when individuals (young adults) cannot rely on help from state institutions but draw on their own resources and take personal responsibility for risk. Beck's ideas are still widely discussed and criticized. For example, one argument is that Beck's definition of risk is predicated on inconsistent "realism" and "constructivism" (Leiss 1994). It has also been asserted that Beck exaggerates the importance of modern risk, since we now live in safer times (Lupton 1999, p. 184). Yet such critical statements do not diminish the importance of Beck's ideas, since his point of view is original and reflexive; it is expressed in terms of the ability to respond to the acute problems of our time.

According to Zubok, youth generally belong to the most risk-prone groups. Innovations are sometimes risky, and youth is a transitional state in the human life cycle. As Zubok notes,

> active interaction (of youth) with new social mediators, its inclusion in the global world, ability to feel, learn and master new emerging social patterns (economic, political, cultural) give it not only dynamism, but also make them the most important driving force for innovative development.
>
> (Zubok et al. 2016, p. 35)

Their innovative potential is in high demand, especially during periods when social and economic systems are developing. When planning the social and economic development of strategic territories, it is particularly important to take into account the position of young people as a special social group that will occupy the leading role in driving this development.

Analysis of regional migration flows in Russia's Arctic has shown a decrease in the population of young people in the majority of regions, with the exceptions being Krasnoyarsk Territory, Yamal-Nenets Autonomous District and Murmansk Region. Most people leave for another region of the Russian Federation, whereby this can be considered mainly an interregional flow (Table 1.1). One in three people leaving the Arctic and one in three entering it is a young person between the ages of 14 and 30. Many young people are leaving their native Arctic towns; on the other hand, however, many are also coming to the Arctic. These flows partially compensate each other in quantitative terms but do not have a significant effect on the resident population in the region, since migration is also high among the elder population.

According to Mkrtychian, most people migrate between the ages of 17 and 19 years graduating from upper secondary school. The main reason for youth migration at this age is the search for educational opportunities and professional self-realization (Mkrtychian 2015).

This introduction to the issues may be summarized in the following observations:

- the Arctic is experiencing a decline in population;
- the Arctic has the most mobile population in Russia;
- youth are underrepresented in the population of the Russian Arctic; and
- youth are the most mobile part of the Arctic population.

These points suggest that measures should be developed for retaining youth in Arctic industrial cities. Indeed, the future development of the Arctic regions depends largely on the young generation's perception of their own prospects and their choice about whether or not to leave these regions. However, research to date has largely overlooked these perceptions and choices, which are the most important considerations for the future of the Russian Arctic

Methods and materials

To answer the question, "Would you prefer to live and work here in the North or to leave?", we conducted a sociological survey as part of a larger research effort studying youth wellbeing in Arctic industrial cities; the study used both qualitative and quantitative approaches. The main goal of the research was to assess perceptions of wellbeing and prospects for the future among young people living in Arctic monocities. The survey sought to identify the general motivation, attitudes and values of youth regarding life, training and work. We invited youth to participate in the survey only in Arctic monocities. Our sample sites were:

Table 1.1 Youth mobility by region in the Russian Arctic, 2018

Region	Youth migration surplus in all migration types (intraregional, interregional and international)	Youth migration surplus in interregional migration	Share of interregional migration in total youth outflow for all types of migration (%)	Share of youth of total number of those migrated interregionally (those who left) (%)	Share of youth of total number of those migrated interregionally (arrivals) (%)
Republic of Karelia*	−141	−135	46.7	40.2	37.6
Komi Republic*	−642	−557	85.7	27.4	30.9
Krasnoyarsk Territory*	594	301	72.7	32.1	37.6
Republic of Sakha (Yakutia)*	−147	−26	28.9	32.0	34.6
Arkhangelsk Region*	−409	−1401	45.1	33.1	36.6
Nenets Autonomous District	−114	−124	47.9	36.3	32.2
Murmansk Region	624	275	63.1	31.7	38.4
Yamalo-Nenets Autonomous District	253	292	65.6	31.7	35.2
Chukotka Autonomous District	−126	−116	78.3	29.0	31.1
Total, Russian Arctic / Average, Russian Arctic	−108	−1491	61.6	32.6	34.9

* For regions whose territory belongs to the Arctic in part, data are indicated only for the Arctic territories of these regions.

- Novy Urengoy, Yamal-Nenets Autonomous District, "the gas capital of Russia", which stands out from other Russian cities by virtue of its sustained economic growth and economic prosperity (Stammler and Eilmsteiner-Saxinger 2010; Men'shikov 2015);
- Neryungri, Republic of Sakha (Yakutia), one of the monocities classified by the government as "undergoing risks of a worsening of the social and economic situation" (Government of the Russian Federation 2014a); and
- Kirovsk, Murmansk Region, characterized as a municipality "having the most difficult social and economic situation" (ibid.).

As the list indicates, Neryungri and Kirovsk are on the negative end of the scale compared to Novy Urengoy in terms of social and economic indicators (see also Ivanova et al., this volume, for a comparison of youth policy implementation in these same cities).

All respondents to the survey were between 14 and 30 years of age, thus falling within the definition of "youth" in Russian law (Government of the Russian Federation 2014b; Kekkonen et al. 2017). Since young people as a social group are heterogeneous, it was necessary to include both studying and working young people, reflecting their particular situation at different stages of growing up and becoming an individual socially and professionally. The lower age limit of 14 years is characterized by the emergence of physical maturity and of rights and social responsibilities. The upper limit of 30 years corresponds to economic independence and professional and personal stability (Kozhurova 2012). The diversity of views and perceptions among young people as an age group has been confirmed in qualitative research carried out by colleagues in the same project Their findings indicate that the same variable can function as an incentive to leave a city among school-age youth, but contribute to its attractiveness as a place to realize one's aims in life for young adults having or wanting to establish a family.

The sample for the survey drew on a stratified random selection, with gender in addition to age employed as principal segmentation factors. A total of 436 people took part in the survey. To understand the full picture of the survey, we present below the age distribution of the youth of the focal towns who took part in the study (Figure 1.2).

Of the total number of respondents, 45.4% were residents of Novy Urengoy, 42% residents of Neryungri, and 12.6 % residents of Kirovsk; that is, the sample is distributed approximately equally between the "crisis" towns and the prosperous town. The reasons for the different classifications are set out at the federal level (Government of the Russian Federation, 2014a).

In addition to the quantitative survey, our results reflect insights from detailed in-depth interviews, focus group discussions and participant observation with young people in all three case cities. The data were complemented by conversations with experts in the municipalities and industrial companies responsible for youth, staff development and social responsibility.

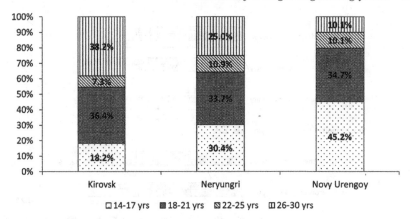

Figure 1.2 Respondents' age distribution in the three case towns (per cent).

Discussion of results

Young people's feelings about migration in the Arctic industrial towns

The results of our survey suggest that the basic values among youth in our Arctic monocities are material security (90.9% of respondents), a strong family and the opportunity to unlock one's personal potential (87.3% of respondents for each). These figures suggest that the extent to which young people see these cities as environments supporting these values will influence any consideration they give to migrating. However, decision-making often happens on a far finer-grained level. In this section we present some findings reflecting that level, our purpose being to develop a clearer understanding of what constitutes wellbeing for youth in an Arctic industrial city. Young people's interest in migrating most often relates to their educational opportunities and professional self-realization. This decision directly affects the life of a young person and the spatial distribution of human capital in general (Kashnickii et al. 2016). Correspondingly, an individual young person's feelings regarding migration will differ depending on the stages of his or her social development.

Most of the youth in the Arctic industrial towns have considered leaving their hometown: 68.5% would like to leave; another 13.5% would like to leave, but thus far have had no opportunity; and only just under one-fifth (18%) plan to stay. If we consider the current situation with reference to the age groups of young people, interest in migration is strongest between the ages of 14 and 21 (Figure 1.3) and most often relates to educational opportunities. According to Mkrtychian, the ages between 15 and 19 represent a peak in migration activity (Mkrtychian 2017, p. 225).

Where education as a reason to leave is concerned, the difference between cities can also be explained by the fact that there is one educational organization in Kirovsk offering secondary vocational education programmes, while Novy Urengoy and Neryungri both have three such organizations. Neryungri has a branch of North Eastern Federal University in Yakutsk where students

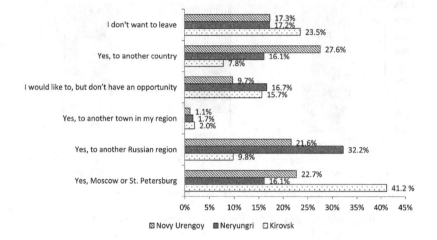

Figure 1.3 Youth attitudes toward migration in Arctic monocities, per cent by monocity, per cent by age group.

can obtain a higher education, for the most part in technical fields. In Novy
Urengoy, branches of universities were closed due to their poor quality and the
poor reputation of the qualifications they award. In Kirovsk, youth do not have
to move to get a higher education, as the neighbouring town Apatity, within half
an hour by bus, has two branches of universities in Murmansk, as well as the
Kola Science Centre of the Russian Academy of Sciences. Apatity has a more
vibrant student life than Kirovsk, and some students want to move there; however, this does not qualify as a significant instance of outmigration, as both
towns are geographically and economically related to each other. Many of the
youth organizations operate in both cities, and young people move back and
forth frequently, sometimes several times a day (see also Bolotova, this volume).

Another salient consideration is that the number of people of reproductive
age is shrinking. This poses a threat to the reproduction of human capital in
Arctic (post-)industrial towns when educational migrants do not return to
their native region after graduation (Andreenkova 2010). We call this the
uncompensated aspect of educational migration. For young people in the
(post-)industrial Arctic towns, large metropolitan areas and even foreign
countries are more attractive than home: such larger communities are mag-
nets in providing extensive educational services as well as a higher standard
of living, even though some respondents were aware that housing and other
services are more expensive there (Figure 1.4).

By the time most young people already have a professional education
(between the ages of 22 and 25), the proportion of those who focus on other
regions of Russia increases, or inclinations to migrate turn into "conscious"
desires, such as "I want to move, but I don't have an opportunity to do so."
In the age range of 26 to 30 years, the share of those who do not want to leave
increases sharply, to approximately one in four respondents.

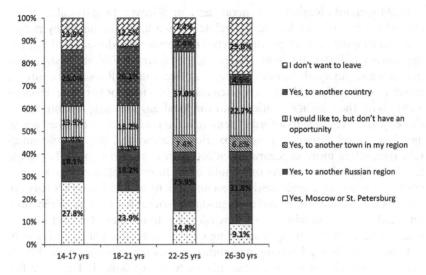

Figure 1.4 Youth attitudes toward migration in Arctic monocities, per cent by monocity, per cent by age group.

Every second young adult notes that in another region there is an opportunity to find an interesting job (49.7%), get a higher salary than in their native region (40.7%), continue their studies and professional development (40.7%) or just to see the world (44.7%).

In light of the qualitative research, some of the differences between the case cities become more easily understood. For example, almost all of our survey respondents in Kirovsk have had contact with foreigners at some point in their life and also travel frequently to Moscow and St. Petersburg. The city positions itself as a tourist centre in addition to being a mining town, and receives both Russian and foreign tourists during the winter season. Moreover, the nearby Kola Science Centre has long had extensive international contacts and visitors, as has the local research station maintained in the mountains near Kirovsk by Moscow State University. Contact with outsiders is something normal and part of everyday life and work. In comparison to Neryungri and Novy Urengoy, Kirovsk also has a much longer history, one starting in the 1920s rather than the 1970s. We hypothesize that the reasons why fewer people want to leave Kirovsk include its less isolated geography and its better-embedded local patriotism. Among those who intend to leave, Moscow and St. Petersburg are the most logical destinations, being only a 90-minute flight or a day's drive away.

Perhaps more frequent travel to Finland and more first-hand experience abroad as individual travellers without tourist or education programmes has also contributed to the youth in Kirovsk being more realistic about the prospects of life in another country. Only 8% of the respondents from Kirovsk aspired to live in another country; the figure for Neryungri was double this, and for Novy Urengoy almost triple.

In Murmansk Region in general, and in Kirovsk in particular, travel abroad is often not for leisure, but related to work or income: many people travel to Finland to work picking berries in summer and do seasonal work in the tourism industry in winter. Moreover, they travel to Finland to shop during sales and buy quality goods at cheaper prices than in Russia. By contrast, travel abroad in Novy Urengoy is often related to leisure or personal development. With their higher salaries from the local gas industry, inhabitants of this city are often able to afford expensive holidays already at a rather young age. They travel abroad in a more luxurious manner and live, eat and entertain themselves more expensively. Indeed, before the crisis of 2014, in the northern Finnish ski resorts one could often hear the phrase "Gazprom is coming", meaning a great deal of revenue was to be had from rich Russian tourists during the winter holiday season. In somewhat of a simplification, one could say that inhabitants of Kirovsk travel to northern Finland to serve inhabitants of Novy Urengoy spending their holidays there at the ski resorts.

In addition, Novy Urengoy has numerous gas company-sponsored partnerships with German cities, especially its twin city, Kassel. There are frequent school exchanges at the secondary level, as well as regular video conferences linking students in Novy Urengoy and Germany. For the most part these are extremely well-organized programmes giving Russian children the opportunity to travel to Germany. In this light, it may be easier to understand why more youth from Novy Urengoy than from Kirovsk might wish to live abroad, since the latter have encountered fewer affluent foreigners at home or abroad.

On balance, we can interpret the survey results as indicating that the leisure-oriented travel abroad of Novy Urengoy inhabitants, along with the schoolchildren's travel and interaction with elite schools sponsored by the gas company, creates an image of living abroad as being easier, more comfortable and more luxurious than in Russia. In comparison, for inhabitants of Kirovsk, living abroad is associated more with hard work and therefore may not be a particularly desirable dream for youth to pursue.

Factors affecting an inclination to migrate and wellbeing among youth

Intentions that young people may have to migrate are formed under the influence of both external and internal factors. In this section, we argue that migratory moods, or intentions, also tell us about young people's senses of wellbeing when living in a northern industrial city: the greater one's sense of wellbeing in one's current residence, the less impetus exists for making the effort to migrate. Young people represent a social group which, due to its high mobility and ambition, may not yet have put down roots in a specific place, making the decision to move easier for them. Aliev considers migration flows the most visible and easily identifiable manifestation of the general level of wellbeing of a town or a region (Aliev 2012, p. 54). Flows go consistently from less prosperous territories to more prosperous ones. According to Russian studies, the main reasons for the outflow of young people from small towns are low incomes, narrow, sector-specific labour markets and other social and economic problems intrinsic to the periphery. However, these disadvantages are tempered by the particular circumstances in the Arctic industrial cities, where the level of income is much higher than in other small cities of the country. Employees in industrial companies can often anticipate receiving a higher salary in the Arctic than they would get for similar work in St. Petersburg or Moscow. This shows that money may not be the main driving force for migration in these cases.

More so than money, life satisfaction is one of the integral and complex indicators for assessing wellbeing. It may include satisfaction with one's position in society, financial situation and other achievements (Andreenkova 2010). In general, much social science research has considered life satisfaction as almost synonymous with wellbeing.

We proceeded to carry out a factor analysis to assess the young people's migration plans. The sociological significance of applying factor analysis is to compress the data set into a matrix that reflects the same information, but with fewer variables. This analysis is based on the premise that variables are indirect manifestations of a small number of latent factors (Nivorozhkina 2008).

Three dimensions of wellbeing: towards defining criteria

We acknowledge that from a qualitative point of view, more in-depth information uncovers the deeper meanings of each of these factors for the youth in our case cities. Most importantly, "satisfaction with wellbeing" is such a broad concept that a limited set of factors cannot possibly grasp the diversity of worldviews and approaches subsumed under it. Cultural research on wellbeing has demonstrated this diversity and identified the lack of clarity of wellbeing as an analytical concept (cf. Stammler and Toivanen, this volume).

Our list of factors is an attempt to categorize some of the indicators that would be related to broader notions of wellbeing in general, including social and economic considerations.

We suggest subsuming youth's satisfaction with their personal economic situation under the larger heading of "financial considerations", which includes the following indicators: income level, living conditions, guarantee of obtaining one's desired job and, as a result, confidence in a prosperous future (Table 1.2). Young people's interest in high salaries and maximum realization of their professional competencies is obvious, since salaries are the main source of income for a modern person, providing most of his or her daily needs: "[Salary] is a measure of society's assessment of the significance, value and prestige of labour" (Konstantinovskii et al. 2013). However, we do not claim that personal finances are the only, or always the principal, factor determining a young person's decision to live in or leave northern industrial cities.

The second group of indicators relates to the social sphere, under which we subsume "satisfaction with leisure and rest", "health", "quality of health care" and "environmental situation". Moreover, "satisfaction with one's

Table 1.2 Understanding wellbeing: from indicators to dimensions

Factors	1	2	3	4
Satisfaction with one's life in general	0.355	0.104	0.766	0.035
Satisfaction with income	0.824	0.197	0.035	−0.091
Satisfaction with standard of living	0.719	0.312	0.153	0.079
Employability\ guarantee of employment	0.86	0.232	−0.0006	−0.127
Certainty in the future	0.728	0.39	0.174	−0.001
Satisfaction with leisure activities	0.173	0.799	0.227	−0.061
Satisfaction with wellbeing	0.261	0.788	0.108	0.124
Satisfaction with the medical care system	0.345	0.708	−0.101	0.017
Satisfaction with ecology	0.259	0.751	0.035	−0.229
Level of happiness	−0.079	0.066	0.873	−0.057
Gender	−0.076	−0.054	−0.029	0.975

Note: Value for Kaiser-Mayer-Olkin sample adequacy after the final selection of observations: 0.858 (more than 0.5)

social surroundings", meaning relatives, neighbours, co-dwellers in the city, peers in school and at work, has an important impact on senses of wellbeing. We argue that a high value for this factor shows young people's confidence in their social wellbeing and sense of security.

We assign the last indicator "emotional satisfaction and level of happiness with one's life" to the mental sphere, on the most personal level.

For young people in Arctic monocities, wellbeing comprises financial, social and mental determinants. Together, these contribute to a sense of security and stability that may encourage an adult to stay in the North. In the present case, a look at the current situation in the three cities shows that young people rely on their own feelings of wellbeing. Clearly, wellbeing in Arctic monocities is also influenced by policy at both the federal and regional levels, the existing and potential conditions for social and economic development of the towns, the attitude of the local authorities, the business climate and the community. However, it is necessary to have a clearer understanding of those individuals on whose wellbeing we are focusing. When young people see their Arctic towns as providing opportunities to meet needs associated with their basic values (satisfaction with income level, certainty about the future, satisfaction with leisure), they may be less inclined to leave.

In light of the above, we suggest that place-based dimensions of wellbeing should be considered in analysing the feelings of youth in Arctic industrial cities regarding migration. Factors influencing such sentiments should serve as the basis for developing recommendations on improving youth policy at the national and regional levels, with the aim being to "keep" young people in the Arctic and thus ensure viable development of the region.

Conclusion

This chapter has outlined a relation between dimensions of wellbeing and intentions to migrate among young people in Russian Arctic industrial towns. The conditions in such industrial cities are often considered less favourable than in other places in the country. We have argued that the combination of five structural factors that influence the development of such cities and are characteristic of young people in general may increase youth outmigration from such cities: the Arctic is experiencing a decline in population, although its strategic importance has increased and industry is still developing there. There are fewer young people in the Arctic than elsewhere, a situation exacerbated by the fact that the Arctic has the most mobile population in Russia, and youth are the most mobile part of the population in general. On the other hand, youth are the most innovative part of the population and are more willing to take risks than other age groups. In this light, given the need for development in the Arctic, there is a corresponding need to keep youth in Arctic industrial cities. We have argued here that if the views of young people on their own wellbeing (satisfaction with income level, certainty about the future, satisfaction with leisure) are not taken into consideration in strategic planning in such cities, in many cases an inclination on their part to migrate may translate into outmigration.

We conducted a survey on youth wellbeing in three Arctic monocities – Kirovsk, Novy Urengoy and Neryungri. Qualitative fieldwork in the cities then validated the observations made on this basis as well as their context-sensitive interpretation.

Our research results show that youth in Arctic industrial towns express an interest in migrating. Yet respondents' narratives, as well as survey data, suggest that young people do not necessarily act on their intentions to migrate, especially when they perceive the conditions in their hometown for their social, financial and mental wellbeing as being favourable. Given the strategic importance of Arctic industry for the Russian economy, the issue of preserving or increasing the population in Arctic cities is particularly relevant. We argue that some of these cities may have a future, even when they run out of the mineral resources they were built to extract, if youth see them as enabling the dimensions of wellbeing better than other locations. Some of the cities, such as Kirovsk, have already started to diversify their economy. Relevant reforms which need active young people for their implementation are being carried out.

A policy of keeping youth in a particular community should be based on the development of existing and the creation of new economic and social incentives geared to enhancing the factors that promote youth wellbeing. If the social environment of the Arctic region continues to become more satisfying and people-centred, Arctic industrial cities can compete favourably with other cities for the best, most innovative and most creative people, who will be crucial in driving development.

References

Aliev, P. R. (2012) 'Vliianie migratsii na trudovoi potentsial molodezhi regiona', *Sovremennye issledovaniia sotsial'nykh problem, 12*, pp. 54–68.

Andreenkova, N.V. (2010) 'Sravnitel'nyi analiz udovletvorennosti zhizn'iu i opredeliaiushchikh ee faktorov', *Monitoring obshchestvennogo mneniia, 5*, pp. 189–215.

Beck, U. (1992) *Risk society: Towards a new modernity*. London: SAGE publications.

Bolotova, A. (this volume) 'Leaving or staying? Youth agency and the livability of industrial towns in the Russian Arctic', in Stammler, F. and Toivanen, R. (eds) *Young people, wellbeing and placemaking in the Arctic*. London: Routledge, pp. 53–76.

Fauzer V. V. (2016) *Demograficheskie i migratsionnye protsessy na rossiiskom Severe: 1980–2000*. Syktyvkar: Izd-vo SGU im. Pitirima Sorokina.

Government of the Russian Federation (2014a) *O perechne monoprofil'nykh munitsipal'nykh obrazovanii Rossiiskoi Federatsii (monogorodov)*. Available at: https://base.garant.ru/70707138/ (Accessed: April 25 2020).

Government of the Russian Federation (2014b) *Osnovy gosudarstvennoi molodezhnoi politiki Rossiiskoi Federatsii na period do 2025 goda*. Available at: http://government.ru/docs/15965/ (Accessed: March 17 2021).

Heleniak, T. (2014) *Migration in the Arctic. Arctic yearbook*. Available at: https://arcticyearbook.com/images/yearbook/2014/Scholarly_Papers/4.Heleniak.pdf (Accessed: April 25 2020).

Ivanova, A., Oglezneva, T. and Stammler, F. (this volume). 'Youth law, policies and their implementation in the Russian Arctic', in Stammler, F. and Toivanen, R. (eds) *Young people, wellbeing and placemaking in the Arctic*. London: Routledge, pp. 147–169.

Kashnickii, I.S., Mkrtychian, N.V. and Leshukov, O.V. (2016) 'Mezhregional'naia migratsiia molodezhi Rossii: Kompleksnyi analiz demograficheskoi statistiki', *Voprosy obrazovaniia, 3*, pp. 169–203.

Kekkonen, A.L., Simakova, A.V. and Stepus', I.S. (2017) 'Osobennosti prostranstvennogo razvitiia i rasshirennoe vosproizvodstvo chelovecheskogo kapitala v Respublike Kareliia i arkticheskikh regionakh Evropeiskogo severa Rossii', *Mir ekonomiki i upravleniia, 17* (2), pp. 86–96.

Konstantinovskii, D.L., Voznesenskaia, E.D. and Cherednichenko, G.A. (2013) 'Rabochaia molodezh' segodnia: Obrazovanie, professiia, sotsial'noe samochuvstvie', *Sotsial'naia nauka i sotsial'naia praktika, 2*, pp. 21–38.

Kozhurova, A.A. (2012) 'Demograficheskoe povedenie molodezhi regionov severo-vostoka Rossii', *Sotsial'naia sfera, 16*, pp. 60–64.

Leiss, W. (1994) *Review of Ulrich Beck. Risk society, towards a new modernity*. Available at: http://www.ualberta.ca/~cjscopy/articles/leiss.html (Accessed: April 25 2020).

Lupton, D. (1999) *Risk*. London: Routledge.

Mart'ianov, V.S. (2013) 'Strategiia gorodskogo razvitiia v Arkticheskom regione Rossii', *EKO, 5*, pp. 124–138.

Men'shikov, A. (2015) 'Gazovaia stolitsa: ot poselka geologov – Do samogo perspektivnogo goroda Iamala', *Rossiiskaia gazeta – spetsial'nyi vypusk, 140*. Available at: https://rg.ru/2015/06/30/reg-urfo/gaz.html. (Accessed: April 25 2020).

Mkrtychian, N.V. (2015) 'Vozrastnaia struktura naseleniia Rossii i ee vliianie na vnutrenniuiu migratsiiu', *Nauchnye trudy: Institut narodnokhoziaistvennogo prognozirovaniia RAN, 13*, pp. 209–221.

Mkrtychian, N.V. (2017) 'Migratsiia molodezhi iz malykh gorodov Rossii', *Monitoring obshchestvennogo mneniia, 1*, pp. 225–242. doi: 10.14515/monitoring.2017.1.15.

Monogoroda Arkticheskoi zony RF: problemy i vozmozhnosti razvitiia: analiticheskii doklad (2016) Available at: http://www.arcticandnorth.ru/Encyclopedia_Arctic/monogoroda_AZRF.pdf (Accessed: April 25 2020).

Nivorozhkina, L.I. (2008) *Mnogomernye statisticheskie metody v ekonomike.* Rostov n/D: Nauka-Spektr.

Russian Federation, President. (2020) *Ukaz Prezidenta RF 'O Strategii razvitiia Arkticheskoi zony Rossiiskoi Federatsii i obespecheniia natsional'noi bezopasnosti na period do 2035 goda'.* Available at: https://www.garant.ru/products/ipo/prime/doc/74710556/ (Accessed: March 2 2021).

Stammler F. and Eilmsteiner-Saxinger G. (2010) *Biography, shift-labour and socialisation in a northern industrial city – The far North: Particularities of labour and human socialisation.* 2nd edited version. Rovaniemi: Arctic Centre. Available at: http://urn.fi/URN:NBN:fi:ula-201111221215

Stammler, F. and Toivanen, R. (this volume). 'The quest for a good life: Contributions from the Arctic towards a theory of wellbeing', in Stammler, F. and Toivanen, R. (eds) *Young people, wellbeing and placemaking in the Arctic.* London: Routledge, pp. 1–13.

Statistics Russia (2018) *Otsenka chislennosti postoiannogo naseleniia sukhoputnykh territorii Arkticheskoi zony Rossii na 1 ianvaria 2018 goda.* Available at: http://www.gks.ru/free_doc/new_site/region_stat/calendar1-2019.htm (Accessed: April 25 2020).

Zubok, U.A., Rostovskaia, T.K. and Smakotina, N.L. (2016) *Molodezh' i molodezhnaia politika v sovremennom rossiiskom obshchestve.* Moscow: ITD «Perspektiva».

2 Not wanting to be "Stuck"

Exploring the role of mobility for young people's wellbeing in Northern Finland

Teresa Komu and Ria-Maria Adams

Introduction

It is common for young people in rural Finland leave their place of birth. Like all the Nordic countries, Finland is experiencing population growth in its larger, urban settlements, and population decline in its smaller, rural ones (Jungsberg et al. 2019, p. 23). Among the Nordic countries, Finland has the largest number of rural regions with declining youth populations, and Lapland is one of the most seriously impacted regions (Karlsdóttir et al. 2020). The trend towards urbanization is largely driven by young people, who account for a significant proportion of Finland's internal migration, and it is peripheral, rural areas, in particular, that are losing young people (Aro 2018; Valtion nuorisoneuvosto 2019). Far from being solely a Finnish phenomenon, the outmigration of rural youth is a global trend affecting the entire Arctic and the developed countries in general (Carson et al. 2016; Farrugia 2016; Corbett and Forsey 2017). Indeed, it has been suggested that as part of growing up in rural places, young people have to "learn to leave" them (Corbett 2007; Kiilakoski 2016, pp. 47–49).

Many researchers have referred to a "mobility imperative" when discussing the outmigration of rural youth: young people are being driven out by structural inequality whereby cultural and economic capital becomes concentrated in cities, leaving rural living symbolically portrayed as inferior to urban life (Armila 2016; Farrugia 2016; Tuuva-Hongisto et al. 2016). Such a view not only casts migration in a predominantly negative light but also risks overlooking young people's own views and agency. Furthermore, research on youth migration has shown that a desire to migrate is not tied to local conditions and that youth outmigration is not a specifically rural phenomenon (Tuhkunen 2007, pp. 10, 145–146). Young people in general are highly mobile and tend to leave rural areas, whether or not these are in decline (Seyfrit et al. 2010; Carson et al. 2016, p. 382).

We suggest that one might approach rural youth outmigration in northern Finland as a manifestation of a "culture of migration" (see Massey et al. 1993; Horváth 2008). In addition, we wish to focus on young people's own understandings of wellbeing and on the kind of role mobility plays in these. We understand migration as an act of individual agency,[1] one driven by

DOI: 10.4324/9781003110019-4

future aspirations that are facilitated and constrained by material, social and cultural conditions (Carling and Collins 2018). We further the discussion started by recent anthropological research on the relations between place and wellbeing (see Ferraro and Barletti 2016), in particular the article by Neil Thin (2016) dealing with the interplay between wellbeing, place and mobility. Thin argues that research and policymaking often assume that wellbeing is rooted in being in a particular place and that certain place characteristics, as well as place attachment, determine a person's wellbeing. Instead, he calls for an understanding of wellbeing as the result of people's dynamic interactions with places, in which purposeful relocations may play an important role in the pursuit of wellbeing (Thin 2016, pp. 6–11).

In the following, we argue that mobility, in the form of both everyday mobility and long-term migration to another place, is necessary for the wellbeing of young people in rural areas, whose lives are characterized by long distances. Further, we suggest that outmigration has come to represent a key transitional phase in young rural Finns' pursuit of individual life paths and self-actualization—components of the "good life" in modern Western societies. To date there has been little discussion between the fields of youth migration and the "anthropology of the good" in the context of the Global North. By combining these perspectives, our research aims to fill gaps in the present knowledge. Next, we discuss how the relations between mobility and wellbeing appear in our case study of young people living in, leaving and returning to their place of birth, Kolari, a municipality in the northern "periphery" of Finland.

Northern culture of migration

Around the world, especially in remote areas, "cultures of migration" have arisen in which youth migration is normalized, valued and expected (Easthope and Gabriel 2008). The current generation of young people is increasingly mobile (Robertson et al. 2018) and is arguably affected by discourses that normalize mobility (Jamieson 2000; Kiilakoski 2016; Rönnlund 2019). According to several prominent authors, Western societies have experienced a fundamental shift in values over the last century and have begun to encourage and value individualization and mobility over social stability (Taylor 1989; Giddens 1991; Bauman 2001). No longer bound to kinship relations and place, people have gone from "living for others" and from prioritizing local communities to pursuing self-actualization and a "life of one's own" (Beck and Beck-Gernsheim 2002, p. 54). Instead of following in their parents' footsteps, people in modern societies are encouraged to pursue individual life paths. Migration may offer a means to realize one's individualistic pursuit of the good life (Tuhkunen 2007, p. 12, 159).

When mobility is the cultural norm, people who stay rooted in their home places become regarded as anomalies. Especially in the rural context "stayers" often find themselves negatively stereotyped as being "left behind" and disadvantaged (Jamieson 2000; Stockdale and Haartsen 2018). According to Timo

Aro (2007a, 2007b), migration from rural and peripheral areas has a long-standing history in Finland. Migration trends in the country have flowed in the same directions since the nineteenth century: from rural areas to urban growth centres, from north to south and from east to west. In the 2000s, a major proportion of those migrating internally consist of young people and students (Aro 2007a, 2007b, p. 375). Outmigration has been a part of life in the remote rural areas of northern and eastern Finland for several generations and has reportedly become a social norm for young people. For them, moving is a way to "realize one's potential", while staying is a sign of being "stuck" (Ollila 2008; Hartikainen 2016). In fact, some scholars have argued that, especially in rural areas, leaving one's birthplace has become part of adolescence (Tuhkunen 2007, p. 159; Robertson et al. 2018).

Throughout the twenty-first century, young people in Finland have expressed a high readiness to migrate to search for employment.[2] A willingness to migrate seems to be highest in rural and peripheral areas, and it is not uncommon for municipalities there to lose 40 to 60 per cent of their young population (Penttinen 2016, p. 149; Karlsdóttir et al. 2020, p. 36). However, "movers" are also the dominant group in urban areas (Tuhkunen 2007, pp. 10, 145–146). A birth cohort study that followed all Finnish children born in 1987 indicates that by the age of 28 only 4 per cent had never moved and more than half (52 per cent) had moved on between four and nine occasions (Moisio et al. 2016, pp. 34–35).[3] When examined by municipal borders, 90 per cent of the area of Finland experiences outmigration (Heleniak 2020, p. 46). Young Finnish people, and Nordic youth in general, also tend to leave their childhood homes 10 to 12 years earlier than youth living in southern Europe (Karlsdóttir et al. 2020, p. 34).[4]

Nevertheless, attitudes towards youth migration are twofold, and young people are both encouraged and criticized for following their individual aspirations (Corbett and Forsey 2017, p. 430). It seems that especially in rural areas youth outmigration causes a conflict between the individual pursuit of wellbeing (young people) and the promotion of collective wellbeing (local communities). Current policy discourses tend to encourage youth mobility and educational aspirations (Robertson et al. 2018). Modern societies need flexible, mobile labour, and young people are encouraged to become educated and to be willing to move to search for work. At the same time, young people are urged to be loyal and to stay to rebuild their local communities, whose viability and social cohesion are seen as threatened by modern individualism and mobility (Beck and Beck-Gernsheim 2002, p. 3; Corbett and Forsey 2017, p. 430). In peripheral and rural areas, educational mobility and youth outmigration prompt concerns, such as "brain drain" (Pitkänen et al. 2017, p. 95), demographic and economic decline (Carson et al. 2016, p. 382) and the concomitant increase in the median age of the population (Hamilton 2010, p. 7). Young people are left with a conundrum: they should leave for their own good and, at the same time, stay for the good of their communities (Corbett and Forsey 2017, p. 440).

Attitudes towards migration and mobility reveal implicit assumptions about their relation to wellbeing and "good" or normative ways to live. Migration has been seen both as beneficial and harmful for the wellbeing of individuals and societies. According to Liisa Malkki (2012), research and common sense generally posit stability and attachment as natural states of being, and mobility and migration as pathologies that demand explanation. In Western thinking, place attachment and being "rooted" has historically been given a positive value, whereas mobility and displacement have been considered sources of ill-being and forms of chaos (Tuhkunen 2007, p. 18; Malkki 2012, pp. 30–41). A prominent cause of the criticism directed at "mobile modernity" has been the assumption that displacements pose a threat to wellbeing (Thin 2016, p. 11). As noted by Aro, in agrarian Finland migration and urbanization were perceived as harmful and problematic, and people who migrated as morally dubious. Historically, migration has been encouraged, but it has also been restricted through various governmental actions (Aro 2007a, 2007b).

In various cultures, ritual displacements and deliberate dis-attachment are, in fact, considered essential for both individual and collective wellbeing (Thin 2016, p. 11).[5] We suggest that the normalization of rural youth outmigration and its positive associations implies that in modern Finnish culture it serves as a necessary transition in a young person's quest for "self-actualization" and pursuit of the good life. The sacrifice that comes with leaving one's birthplace and social network is necessary if a young person is to be able to fulfil the modern demand of individualism and of realizing his or her "potential". In turn, those who never leave their birthplace "fail" to meet these expectations, which could explain why stayers are disparaged even if their decision might benefit the community. In sum, in a culture of migration, the wellbeing and self-fulfillment of an individual require mobility.

Methodology

This research is based on ethnographic fieldwork in the municipality of Kolari in Finnish Lapland. All data were collected during multiple fieldwork periods between February 2019 and August 2020. The definition of "youth" varies from culture to culture (Clark 2011). Here, we apply the national, legal definitions of the Finnish government. The 2017 Finnish Youth Act defines persons up to and including the age of 29 years as youths. Our research participants are between 16 and 29 years of age. Participants were informed throughout the research process that our aim was to examine different aspects of wellbeing from the young people's own perspective and consider how they figure in their decisions to either stay or leave Kolari.

Our main research methods were participant observation (Bernard 2006; Robben and Sluka 2012) and unstructured interviews; the latter were conducted in the form of conversations in which interaction between the participants was encouraged. (Sardan 2015). Participant observation included involvement in local family sports events, the local circus's spring show, *penkkarit*—a traditional Finnish event on upper secondary school students' last day of school before starting their studying break—and several visits to the local upper secondary school and youth centre. Furthermore, time was spent at the "hot spots" of the region's winter tourism centres, where young people work and spend their leisure time. In addition, all the shops, "hangouts" and restaurants favoured by local youth were visited. During times of physical absence, social media accounts, such as Instagram and Facebook, were used to "stay connected to the field".

Most of the research material consists of unstructured interviews with young people. The interviews adhered to specific, prepared questions and were conducted while participating in everyday activities of the youth. These conversations lasted from a few minutes to two-hour in-depth conversations that included biographical stories. In addition, two focus group interviews were conducted in two different settings with between three and six participants. Such an arrangement offers a forum in which young people are stimulated by each other's comments, which has advantages for understanding peer-to-peer influences and shared understandings (Clark 2011, p. 72). Furthermore, one in-depth interview was conducted in the neighbouring municipality of Rovaniemi, with an informant who had recently left Kolari for educational reasons and to experience living in a larger city. Fieldnotes were recorded on the participant observation as well as the unstructured interviews.

We received additional information from institutional youth workers and teachers who, by virtue of their profession, provided valuable insights. They served as key persons in initiating contacts with the local youth and made it possible for the research to take place in a setting that was familiar to our informants. During data analysis, the fieldnotes were transcribed, coded and categorized and then analysed in the light of existing research. The quotations that we use in our analysis are taken from the fieldwork notes and the interview transcriptions. Because we are dealing with a small community, all of the names in the quotations are pseudonyms, and we have chosen not to use specific information regarding the names of particular places or the exact ages of our informants.

The municipality of Kolari and the Finnish "periphery"

Finland is one of the Nordic welfare states, whose policies support individual pursuits of self-actualization and wellbeing. Such welfare countries are characterized by comprehensive social security, free education and a high level of social trust. Nordic countries score well in the international rankings of human development, wellbeing, quality of life and happiness (Lundgren and Cuadrado 2020, pp. 130, 138).[6] Finland is a sparsely populated country, most of whose area is classified as "predominantly rural"; it is characterized by strong centralization of the population in southern growth centres (Nilsson and Jokinen 2020, p. 17). Around 5 per cent of the population in Finland lives in remote rural areas, which make up 70 per cent of the country's surface area. By contrast, over 70 per cent of the population lives in urban areas

(Helminen et al. 2020). Generally speaking, rural areas in Finland are experiencing a decline in services, a decreasing and ageing population and outmigration (Sireni et al. 2017, pp. 46, 14–15).

The lives of young rural people are often affected by a lack of transportation, educational opportunities and leisure activities (Wrede-Jäntti 2020, pp. 7, 13), and this is also true of Kolari. Long distances and their effect on everyday life in remote, sparsely populated areas is well recognized in previous research on rural youth (see, for example, Armila et al. 2016). In the case of Kolari, long distances influenced various aspects of the young people's lives from education to shopping, leisure and medical care. Many of the young people whom we interviewed lived either quite long distances from the municipal centre or school or had to commute long distances to see their friends or relatives.

Educational and regional policy decisions have contributed to the outmigration of rural youth. Peripheral areas are losing educational institutions as public services and leisure activities are being concentrated in growth centres (Kivijärvi and Peltola 2016, pp. 5–6). On the other hand, favourable support structures and educational support policies, such as study grants and youth housing, allow young people to become financially independent from their families quickly, supporting them in leaving the parental home (Karlsdóttir et al. 2020, p. 34). Thus, in Finland, while there are young people who wish to move but lack the resources to do so, finances are not as critical a consideration as, for example, in the Russian context (see Bolotova, this volume). This is an example of how welfare politics can encourage and support certain ways of living and how welfare is one way of ensuring the creation of educated, self-managing citizens needed by modern societies (Gulløv 2011, p. 31).

Our case municipality, Kolari, is located in northern Finland close to the Swedish–Finnish border and approximately 120 kilometres north of the Arctic Circle. It is one of Finland's sparsely populated, northern rural municipalities characterized by a narrow economic structure, low intensity of land use, a harsh climate, long distances and remoteness from shopping centres, health care services and urban settlements (Helminen et al. 2014; Borges 2020, p. 73). Kolari is a relatively small municipality; its population of 3,834 and population density of only 1.5 inhabitants per square kilometre (Kuntaliitto 2019a) are characteristic of northern rural towns. In 2018, only 7.2 per cent of the population in the municipality were aged between 15 and 24 (Kuntaliitto 2019b). At the time the fieldwork was conducted, Kolari had 40 upper secondary students, a number reflecting the population dynamics of many northern municipalities, which tend to have many pensioners and few young families and many of whose young people have already left to pursue an education.

Tourism is the most important and a growing livelihood in Kolari,[7] which boasts the ski resort Hiihtokeskus IsoYlläs[8] on Yllästunturi Fell and the popular Pallas-Yllästunturi National Park[9], as well as the thriving tourism village of Äkäslompolo. During the high season in the winter months, thousands of international and national tourists visit the fell villages without ever passing through or visiting the actual centre of Kolari, where the schools, town hall, library, and youth centre are located. The tourist-oriented villages have their own shops as well as a variety of restaurants and bars, which are busy during the tourism season. Many temporary workers arrive from southern parts of Finland during the winter season, but once the season is over, the villages become silent.

While the municipality's tourism villages have a growing number of inhabitants, the overall population and economy in Kolari is declining (Similä and Jokinen 2018, p. 155). Employment and youth outmigration continue to present challenges to the municipality, which is seeking new ways to employ people year-round. For example, since 2006 two different mining companies have been making plans to re-open the old open-pit Hannukainen mine. Although the local tourism industry has protested against it, the mine is expected to bring new job opportunities, boost the regional and local economy and breathe new life into the municipality (Similä and Jokinen 2018; Komu 2019).

Leaving to pursue one's dreams

The pursuit of individual happiness is highly valued in Western cultures (Thin 2009, pp. 30, 35) and this is reflected in our young informants' reasons for moving away from Kolari, where they were born. It is well established in the literature that a significant driver of rural youth outmigration is a desire to find employment and education opportunities (see, for example, Tuuva-Hongisto et al. 2016). However, in the case of Kolari it is not a lack of job opportunities per se that has been driving young people to leave. In fact, young people indicated that they knew there was work available in Kolari in certain sectors, such as tourism or elderly care. The nearby Swedish border makes commuting across national borders an attractive career opportunity, providing higher salaries in some sectors, for example health care. The key motivator for young people, however, was to have the opportunity to do work that would be worthwhile and *meaningful* from their perspective, as the following excerpt shows:

> I am interested in interior design or economic studies and not in elderly care. I know I could get a job immediately in the caregiving sector, but I just don't want that.[10]
>
> (young female)

The participants in our case study area are not primarily seeking temporary employment (*kausityöntekijä*), readily available in the tourism sector. Needless to say, they take jobs to earn some money by cleaning cottages or working at a local enterprise, but in the long run they generally dream of having work that corresponds to their education and aptitudes and is more permanent. Often, young people from southern Finland come to work during the winter tourism season and, once the work is done, they return to their homes. To be sure, some of the local youth dream of being part of the growing local tourism business, in which they see enormous potential for finding future employment.

While the planned mining project could offer future work opportunities and breathe new life into the municipality, young people remained ambivalent toward it. The positive effects of an active mine are visible in the neighbouring town, Pajala, on the Swedish side of the border. A better infrastructure and range of local services—direct benefits of the mine—are shaping the community into a more attractive place to live. On the other hand, the young people in Kolari are unsure if and when exactly the Hannukainen mine would open and what specific jobs it would provide for them. Even though the tensions around the mining project overall are high in the municipality, it did not seem like a relevant issue for many of the young people. As one noted:

> I don't really care about the mining project. Who knows if it really even ever opens? I don't really have an opinion. There is so much for and against the mine. I just stay out of the discussion.
>
> (young female)

Overall, the mining project did not seem like a big attraction compelling young people to stay in Kolari. Again, the *types* of jobs it could provide was a more important factor than its overall ability to provide jobs. Our preliminary findings are in line with previous research, which has demonstrated that even though large-scale industrial projects may create new job opportunities locally, they do little to stop young people from moving away from rural communities (Seyfrit et al. 2010).

An important aspect of the good life and wellbeing for young people in Kolari is being able to work in a job that they consider both meaningful and sustaining (see also, Myllyniemi 2016, p. 27). What makes life meaningful is understood in different ways at the individual level, but it is also conditioned culturally. A cross-cultural study by Gordon Mathews on Hong Kong, Japan and the United States found that in all three societies family and work were the most common answers participants gave when asked: "What makes life worth living?" (Mathews 2009, pp. 176–177, 180). While the opinions on what kind of work was considered meaningful varied individually in our case study, most young people shared a view of work as something that should bring meaningfulness to one's life and provide an outlet for self-fulfillment.

The selectiveness regarding job aspirations observed in our case study seems to be an expression of the previously discussed modern outlook that encourages the pursuit of individual life paths and dreams. An important difference between the Finnish migration of the past and the present is that while migration has perhaps always been sparked by the pursuit of a "better" life, today it is motivated more by personal ambitions and visions of wellbeing than by a compelling need to make a living (Aro 2007a, 2007b, p. 371). This is largely a reflection of the conditions of life in the current welfare state. For example, for the generation born in the beginning of the twentieth century, who experienced war and the rebuilding of the country, chasing a personal "dream job" was probably not a priority amidst the struggles of everyday life (Häkkinen and Salasuo 2016, p. 185).

In pursuing one's dreams, the choice of the right education path becomes an important factor. Lack of secondary education is statistically the strongest indicator for future marginalization and ill-being for young people (Armila 2016, p. 11; Ristikari et al. 2016, p. 97). As in many peripheral municipalities, only upper secondary education is possible in Kolari. For anything else—vocational schools, universities or universities of applied sciences—young people need to move elsewhere. The municipality tries to actively strengthen different educational paths that would not require young people to move away by offering local apprenticeships and distance learning opportunities. Yet after upper secondary school there are only a few options in distance learning classes, causing young people to leave the community to pursue their desired education. Some of the youth already leave after graduating from comprehensive school to attend vocational school or an upper secondary school with special emphasis (e.g., languages, music or art). As the excerpt below indicates, many wished to continue to visit their hometown even after moving away to maintain their relations with family and friends:

> I don't want to leave Kolari except for my education I have to go to Kittilä during school weeks. Whenever it is possible, I return back home.
>
> (young female)

While adolescents (youth under 18 years) tend to still be in the process of deciding whether or not to stay or leave, some of our older informants had already undergone the process of leaving and returning. As noted by Stockdale and Haartsen (2018, p. 2), leaving or staying is indeed not a one-off decision, but one made multiple times over the course of a person's life. The reasons for these decisions vary individually, but they often have to do with returning to "one's roots". Given that remote regions, such as Kolari, are often unable to provide a variety of educational options and services, it has been suggested that the outmigration of young people could be perceived as beneficial, rather than harmful, in supporting their viability in the long run. Instead of trying to stop young people from migrating when leaving is the "social norm", these regions could concentrate on encouraging them to return after they have acquired qualifications that could be much needed in the region (Borges 2020, p. 74).

The perks and perils of "everyone knows everyone"

Anthropological studies on the "good life" generally emphasize the importance of the social dimension for wellbeing (Mathews and Izquierdo 2009). Social relations, instead of place, can act as a grounding and defining factor in one's decision to either leave or stay (Kivijärvi and Peltola 2016, p. 7). While social interaction is a source of wellbeing, individual wellbeing may also be constrained by the dynamics of social life. Kolari is a small community where, as the young people describe it, "everyone knows everyone". Depending on who was being asked, this close connection with peers could be perceived as either a positive or a negative factor.

For some young people, close connections and knowing each other well were connected to feelings of security and familiarity. In some cases, the desire to find a partner in their own community or another "northern" community was an important criterion, as only people from the North would understand their "northern way of life". The importance of maintaining close social relations and their connection to this particular place was the most important factor of the "good life" for some youth, as the following excerpt shows:

> I moved to Tornio during my secondary education, but I returned almost every weekend to see my friends and to hang out at home. I could not find work matching my training, so I decided to retrain for another job so I could stay here and wouldn't have to move far away. For me it is important to be able to stay where my friends and family are.
>
> (young male)

In order to sustain close relations and friendships, young people in Kolari long for more public meeting places and hangouts. They often meet in private locations, such as their homes or cottages, because there is a lack of suitable places in town to gather without being supervised. Being just across the Swedish border, young people regularly visit Pajala for shopping and other activities. However, young people understand that from an economic standpoint it does not make sense to run meeting places in a small town. For adolescents, the municipality has organized a meeting place in the form of a youth centre called "Laguuni", which is very popular especially among adolescents. Professional guidance is provided by social and youth workers, who have a warm-hearted relationship with the local youth. Young people come there to talk about issues that concern them, play billiards, have a coffee or some snacks, play in the newly added "gaming room" on computers (that some of them would not have access to in their homes) or use it as a space for personal regeneration and leisure. These youth centres are essential institutions in enhancing young people's wellbeing in remote towns and are highly valued among younger local youth.

Mobility and, on the other hand, immobility affected social relations in various ways. Living in a culture of migration shapes the way young people learn to conceive the nature of relationships and their temporary nature (Kiilakoski 2016, pp. 37–45). Many of the young people still currently living in Kolari mentioned family, friends and work colleagues who had left to work and live elsewhere. The absence of important people is a recurrent theme in the conversations, and "missing friends and family" was mentioned as a negative aspect of living in a remote place. Yet people who had already migrated could serve as an anchoring factor in prospective migrants' decision to leave. Young people wanted to move to cities where they already "know someone", wanting and needing a contact person (often a relative or a friend) whom they could trust and spend time with.

When it comes to local mobility, the story of one young female participant exemplifies the lived reality of local youth regarding the role of transportation and the importance of the ability to be mobile for upkeeping social relations:

> The only means of transport that I can use on my own is the school bus. I live around 25 km from the centre of Kolari and if I want to hang out with a friend, I have to pre-arrange everything and stay over for the night if my parents are not willing to pick me up. Also, if a friend wants to visit me, she or he has to join me on the school bus, which the friend would have to pay for, which makes things more complicated. Also, I have friends who live on the other side of Kolari and I need to travel 50 km one way if I want to visit them. If I want to do any free-time activities I have to stay here after school and arrange for someone to pick me up afterwards.
>
> (young female)

Social relations often played a role if one decided to return to Kolari after moving away. Starting a family on their own certainly brings some young people "back to their roots", where the social network of a family plays a vital role. While youth migration flows from rural areas to growth centres, migration of parents with children generally flows from cities to rural and semi-urban areas (Moisio et al. 2016, p. 29). Finnish people in general perceive rural areas with ready access to nature as good places to live and raise children (Sireni et al. 2017, p. 45). Leaving may enable a young person to "self-actualize", to get an education and possibly start a family, after which they are "ready" to move back. The proximity to a partner or to family plays an essential role in staying, leaving and returning, as this excerpt demonstrates:

> I left when I was young and stayed many years in southern Finnish cities. But once I got children of my own, I started missing the network of my family. I married someone who is also from here and after a while we decided to move back. Now we both have jobs here and are happy that our kids can grow up here as well. I had a great relationship with my grandparents, and I want the same for my kids.
>
> (young female)

Then again, for some, the narrow social circles of their home area were a reason to leave. Some young people have an urge to experience "something new" and to get to know new people, which, in a peripheral place, where new people only arrive occasionally, is rather impossible. Moreover, as our informants noted, finding a suitable partner is not always easy if "everyone knows everyone" and options are limited.

The importance of having "something to do"

In the participants' perception, having access to a variety of free-time activities, especially winter sports and other outdoor activities, is an important factor for wellbeing in Kolari. On a general level, the informants were satisfied with the leisure activities on offer. The positive examples mentioned include various sport possibilities, which range from downhill skiing to skidoo-driving, and from active soccer, hunting and ice-hockey clubs to the possibility to perform circus art. In addition, there are possibilities to attend a variety of music classes. Some young people also actively engaged in sports activities outside of their municipality borders. For example, some dance classes were provided in the neighbouring municipality of Kittilä, while the circus activities and soccer attracted the youth in Kolari. However, the main consideration was whether transportation could be organized enabling the young people to take advantage of these options as one participant said:

> This is the only place we have been living at so far, so we can't really miss something that we have never had. I am satisfied with what we have, and we have plenty of activities to choose from. The question is how we can organize transportation to and from the places.

As the above excerpt shows, the issue of transportation came up often when the young people described their everyday activities. If young people want to go to the movies, for example, they need to drive about two hours to Rovaniemi. Interestingly, young people do not regard the distances as such as an issue if there is a way get where they want to go. There is a clear distinction between those who are dependent on other people to drive them and those who own a vehicle. The young people who have access to a vehicle did not regard the distances as a problem, while those without access to a vehicle more often had feelings of "being stuck". The issues of insufficient public transportation were raised throughout conversations and interviews as a decisive element of living in a rural area. Young people value independence and want to be able to move between places on their own. Having to rely on a what is a scanty public transportation system creates the negative feeling of being "dependent" on other people.

According to the Finnish Driving License Act (Ajokorttilaki 286/2011) (Finlex 2021) young people can get a driver's license as early as the age of 17 under "special circumstances", which include living in rural places. This law has been well received by local youth and it is applicable to almost every young person living in the Kolari area. However, getting a driver's license is also a financial consideration and not everyone can afford to get a license and a car. For some, the current transportation, which is inadequate, hampers their lives to the extent that they want to move to a bigger city, as explained by a young man who had moved to Rovaniemi:

I used to live quite far outside of Kolari's city centre, literally in the middle of nowhere. Of course, there were outdoor activities but otherwise it was so boring. All my friends lived far away, and it was hard to get to their place. My parents or a friend's parents needed to drive us kids, which was not too easy. Sometimes we travelled with the school buses, stayed overnight at the friend's place and went back to school together on the next day. But it was not an easy way to get around.

What sets living close to a tourist destination apart from other rural places is the variety of restaurants and bars the tourism area offers. However, these are often expensive locations that young people do not find particularly attractive or affordable. During the winter season, several famous musicians and artists come to the tourism villages, but for the local youth the issues of cost and transport are still an overriding factor. The tourist area is located around 35 km from the centre of Kolari, with no public transportation allowing young people to travel back and forth between their homes and the tourism hotspots.

"Something to do" was often the answer when we asked young people what they want to have in their municipality. Even though they expressed satisfaction overall with the free-time activities offered, they said that some additional sports activities, such as dance classes, a swimming hall or a frisbee golf course, would add to their happiness. In their narratives young people noted wistfully that there had previously been more shops and places to hang out in town, such as a kiosk and a pizzeria, but that these had closed because of financial difficulties. Then again, young people understand why it is uneconomical to run a business in Kolari given the small population, especially the number of young residents.

A "dead place" and a place of "beloved nature"

While some young people view Kolari as "a dead place with only little to do", others view it as a safe environment with "great access to nature activities". According to Anne Ollila (2008), young people's migration tendencies in Finland are shaped by discourses that associate rural areas with impending marginalization and urban settlements with success. But "peripheries" have a number of meanings. For example, Finnish rural areas are associated with closeness to nature and safety, but also with conservatism and stagnation. Urban areas are associated with individuality and endless possibilities, but also with loneliness and insecurity (for the full list, see Tuhkunen 2007, p. 155). These perceptions were shared by our informants.

Studies have shown that rural youth may have strong bonds to their home places and often enjoy their rural lives, especially the close connection to nature, regardless of their decision to move away (Armila et al. 2016; Farrugia 2016; Kiilakoski 2016; Penttinen 2016; Tuuva-Hongisto, Pöysä and Armila 2016; Rönnlund 2019, p. 2). A connection to nature plays a big role in the youth's personal wellbeing in our study area and young people described how important it was to just have access to nature, even if they did not always actively make use of it. For some people, access to the specific kind of nature found in the North was the reason to return to Kolari. During the fieldwork we heard extensive descriptions of the surrounding nature: the calming aspect of nature, the beauty, magic, roughness and harsh weather conditions. Nature was described as an energizing and nurturing place. It was the connection to nature that young people missed if they had lived elsewhere. They described how the forests in the southern parts of the country were different and how the feeling of space was lacking when compared to their rural home area, as this young man pointed out:

> The forest is different in the south. I don't know how to explain it but it just feels very different than here. The trees are different and so is the sense of space.

Nevertheless, researchers have pointed out that young people do not leave their home regions only because they are forced to do so, but also because of dissatisfaction with their current rural lives and their longing for city life (Pedersen and Gram 2018, p. 630). Some of our research participants said that their longing for "something different" is so strong that as soon as they could, they would move away. One young man said:

> Mainly I wanted to get away because there was not much to do ... it's a dead place. Did you see the centre? There used to be a few more shops but everything is closed now, except for the two grocery stores and one burger grill.

Urban lifestyle itself can be perceived as something worth pursuing, especially by young people (Penttinen 2016, pp. 149–150). David Farrugia has argued that with their opportunities for consumption and leisure experiences, cities may attract young people by being symbols of modern life and youth culture. Compared to the perceived possibilities offered by cities, rural places may come to be seen as places where "nothing happens" (Farrugia 2016). The experience of the good life is also to some degree relative and results from a comparison between what one has and what others have. Research on stress, for example, has shown that feeling that one does not have enough or *as much as others* is as a major psycho-social stressor (Sapolsky 2004, pp. 372–374). In our case study, dissatisfaction with rural life seems, at times, to have arisen from comparing one's life to what life was imagined to be like elsewhere.

If one's hometown is perceived as a "dead place", moving away is essential to ensure a future for oneself. Migration may offer a way to escape marginal conditions and to actively better one's life. The young people in our case study had clear visions or a direction for their life goals, informing their decision to either stay or leave. In this respect, youth outmigration is an expression of the ability to dream of a better future (Fischer 2014) as well as of the feeling of being able to control the direction of one's life, both of which are important aspects of one's sense of wellbeing (Jankowiak 2009).

Conclusions

In this chapter we have been exploring the role of mobility in young people's wellbeing in northern Finland. Our results suggest that the ability to be mobile is necessary for the wellbeing of young people in rural areas, whose lives are characterized by long distances. The importance of mobility is highlighted also when its counterpart, the inability or refusal to move, results in feelings and accusations of being "stuck". It was common that to access the things they perceived to be most important for their wellbeing—social relations, meaningful activities and work—the young people interviewed needed to be mobile either within or between localities. It is important to point out that for our informants it was not long distances per se that were perceived as a problem, but access to the transportation needed to travel those distances. While previous research on rural youth has discussed the role of long distances in young people's everyday lives, this research is the first attempt to spell out how mobility is connected to young people's wellbeing in sparsely populated rural areas.

In light of our findings, we would also argue that the phenomenon of Finnish rural youth outmigration can be understood as an example of a prevailing culture of migration. In this context, a young person "needs" to leave his or her birthplace to be able to fully pursue the good life in modern Western society, which values individualism and mobility. Discussions of rural youth outmigration often deal with the clash between individual pursuit of wellbeing and the viability and wellbeing of the rural communities affected. Our decision to focus on young people's viewpoints has shown that while rural youth outmigration may have negative impacts on the community level, on the individual level it may be a way for a young person to better his or her life. Migration may offer the means for a young person to pursue his or her dreams or to escape marginal conditions. At the same time, it is important to note that some of the young people interviewed in the present study dreamed of being able to stay in their home region. For them too, mobility could offer the means to realize this dream by leaving their birthplace temporarily to get qualifications that would allow them to come back. While the young people under 18 years often wanted to leave, returning seemed to be the most attractive option for young families and the "older" youth who had already acquired an education.

Account of our study, we have seen that the same considerations can serve as push and pull factors, with no straightforward causation between place characteristics and wellbeing. Our findings support the argument of Thin (2016) that wellbeing should be understood as the result of people's dynamic interactions with places rather than of people having access to certain place characteristics. We extend the argument to the context of rural youth outmigration and demonstrate the key role that various mobilities play in the lives of young people in rural areas. With this chapter we hope to have done justice to young people's multivocal views on wellbeing and the importance of mobility in northern Finland.

Acknowledgements

This publication was supported by the WOLLIE project, funded by the Academy of Finland, decision number 314471. This publication was partly supported by the University of Vienna, funded by the 'uni:docs fellowship' programme.

Notes

1 Our definition reflects the highly individualistic culture in Finland. In some other cultural and geographical settings, migration is understood more as a collective decision, as discussed by Alla Bolotova in her chapter on Russia in this volume.
2 In 2016, more than three of four respondents on a survey were prepared to move to get a job, while nearly half would be fully ready to do so (Valtion nuorisoneuvosto 2016). That same year, among the EU countries Finland had the third highest number of young working-aged people to move within the country for a job (Palen and Lien 2018). Another survey reported that at the time the research was done, 81 per cent of the respondents, young Finnish people, were planning to move to another region (Tuhkunen 2007, pp. 10, 145–146).
3 Migration was most frequent between the ages of 15 and 21 (Moisio et al. 2016, pp. 34–35).
4 Currently, the average age at which children leave the parental home in Finland is 22 years (Karlsdóttir, Heleniak and Kull, 2020, p. 34).
5 As examples Thin mentions long-term ascetic withdrawals and the Indian custom whereby newlywed women are ritually assisted to leave their old homes and to embrace their new ones (2016, p. 11).
6 These rankings are based on measurements such as life expectancy at birth, education and gross national income per capita.
7 Forty-eight per cent of the municipality's economy and 40 per cent of employment came from tourism in 2011 (Matkailun tutkimus- ja koulutusinstituutti 2013).
8 It has the fourth-largest annual revenue of all the ski resorts in Finland (Jänkälä 2019).
9 The park, which recorded 561,200 visitors in 2019, is the most popular national park in Finland (Metsähallitus 2019).
10 The excerpts have been translated by the authors from Finnish.

References

Finlex (2021) *Ajokorttilaki 286/2011*. Avaivable at: https://www.finlex.fi/fi/laki/ajantasa/2011/20110386.
Armila, P. (2016) 'Johdatus: Tervetuloa Hylkysyrjään', in Armila, P., Halonen, T., and Käyhkö, M. (eds) *Reunamerkintöjä Hylkysyrjästä. Nuorten elämänraameja ja tulevaisuudenkuvia harvaanasutulla maaseudulla*. Helsinki: Nuorisotutkimusseura (Nuorisotutkimusseura, verkkojulkaisuja 117, Liike), pp. 9–12. Available at: https://

www.nuorisotutkimusseura.fi/images/kuvat/verkkojulkaisut/reunamerkintoja_hylkysyrjasta.pdf.

Armila, P., Halonen, T. and Käyhkö, M. (eds) (2016) *Reunamerkintöjä Hylkysyrjästä. Nuorten elämänraameja ja tulevaisuudenkuvia harvaanasutulla maaseudulla.* Helsinki: Nuorisotutkimusseura (Nuorisotutkimusseura, verkkojulkaisuja 117, Liike). Available at: https://www.nuorisotutkimusseura.fi/images/kuvat/verkkojulkaisut/reunamerkintoja_hylkysyrjasta.pdf.

Aro, T. (2007a) 'Maassamuuton suuri linja 1880-luvulta lähtien: Savotta-Suomesta Rannikko-Suomeen', *Kuntapuntari, 14*(2), pp. 9–12. Available at: https://www.stat.fi/artikkelit/2007/art_2007-07-13_001.html.

Aro, T. (2007b) 'Valikoiva muuttoliike osana pitkän aikavälin maassamuuttokehitystä', *Yhteiskuntapolitiikka, 72*(4), pp. 371–379. Available at: https://www.julkari.fi/bitstream/handle/10024/100886/074aro.pdf?sequence=1&isAllowed=y.

Aro, T. (2018) *Nuorten aikuisten (25–34 -vuotiaat) ja muun aikuisväestön (35–54 -vuotiaat) korkea-asteen tutkinnon suorittaneiden nettomuutto 2014-2016.* Helsinki: Statistics Finland.

Bauman, Z. (2001) *The individualized society.* Cambridge: Polity Press.

Beck, U. and Beck-Gernsheim, E. (2002) *Individualization: Institutionalized individualism and its social and political consequences.* London: SAGE (Theory, culture & society).

Bernard, H. R. (2006) *Research methods in anthropology: Qualitative and quantitative approaches.* edn 4th. Lanham: Altamira Press.

Bolotova, A. (this volume) 'Leaving or living? Youth agency and the liveability of industrial towns in the Russian Arctic', in Stammler, F. and Toivanen, R. (eds) *Young People, wellbeing and placemaking in the Arctic.* London: Routledge, pp. 53–76.

Borges, L. A. (2020) 'Geographies of labour', in Grunfelder, J., Norlén, G., Randall, L., and Sánchez Gassen, N. (eds) *State of the Nordic Region 2020.* Copenhagen: Nordic Council of Ministers (Nord, 2020:001), pp. 66–75. Available at: https://pub.norden.org/nord2020-001/nord2020-001.pdf.

Carling, J. and Collins, F. (2018) 'Aspiration, desire and drivers of migration', *Journal of Ethnic and Migration Studies.* Routledge, *44*(6), pp. 909–926. doi: 10.1080/1369183X.2017.1384134.

Carson, D. B., Carson, D. A., Porter, R., Ahlin, C. Y. and Sköld, P. (2016) 'Decline, adaptation or transformation: New perspectives on demographic change in resource peripheries in Australia and Sweden', *Comparative Population Studies, 41*(3–4), pp. 379–406.

Clark, C. D. (2011) *In a younger voice: Doing child-centred qualitative research.* Oxford University Press, USA.

Corbett, M. (2007) *Learning to leave – The irony of schooling in a coastal community.* Halifax: Fernwood Publishing. Available at: https://fernwoodpublishing.ca/book/learning-to-leave.

Corbett, M. and Forsey, M. (2017) 'Rural youth out-migration and education: challenges to aspirations discourse in mobile modernity', *Discourse: Studies in the Cultural Politics of Education.* Routledge, *38*(3), pp. 429–444. doi: 10.1080/01596306.2017.1308456.

Easthope, H. and Gabriel, M. (2008) 'Turbulent lives: Exploring the cultural meaning of regional youth migration', *Geographical Research.* John Wiley & Sons, Ltd, *46*(2), pp. 172–182. doi: 10.1111/j.1745-5871.2008.00508.x@10.1111/1745-5871.geographies-of-migration.

Farrugia, D. (2016) 'The mobility imperative for rural youth: the structural, symbolic and non-representational dimensions rural youth mobilities', *Journal of Youth Studies*. Routledge, *19*(6), pp. 836–851. doi: 10.1080/13676261.2015.1112886.

Ferraro, E. and Barletti, J. P. S. (2016) 'Placing wellbeing: Anthropological perspectives on wellbeing and place', *Anthropology in Action*. Berghahn Journals, *23*(3), pp. 1–5. doi: 10.3167/aia.2016.230301.

Fischer, E. F. (2014) *The good life: Aspiration, dignity, and the anthropology of wellbeing*. Stanford: Stanford University Press.

Giddens, A. (1991) *Modernity and self-identity: Self and society in the late modern age*. Cambridge: Polity Press.

Gulløv, E. (2011) 'Welfare and self care: Institutionalized visions for a good life in Danish day-care centres', *Anthropology in Action*. Berghahn Journals, *18*(3), pp. 21–32. doi: 10.3167/aia.2011.180303.

Häkkinen, A. and Salasuo, M. (2016) 'Aika näyttää. Nuoret, hyvän elämän määrittelyt ja sukupolvien katkeilevat ketjut', in Myllyniemi, S. (ed.) *Katse tulevaisuudessa. Nuorisobarometri 2016*. Helsinki: Grano Oy (Nuorisotutkimusverkoston julkaisut), pp. 177–191. Available at: https://tietoanuorista.fi/wp-content/uploads/2017/03/Nuorisobarometri_2016_WEB.pdf.

Hamilton, L. C. (2010) 'Footprints: Demographic effects of out-migration', in Huskey, L. and Southcott, C. (eds) *Migration in the Circumpolar North: issues and contexts*. Edmonton, Alberta: CCI Press, pp. 1–14.

Hartikainen, E. (2016) 'Keskustelun keskellä syrjä: Syrjäseutukeskustelu Suomessa vuodesta 1945', in Armila, P., Halonen, T. and Käyhkö, M. (eds) *Reunamerkintöjä Hylkysyrjästä. Nuorten elämänraameja ja tulevaisuudenkuvia harvaanasutulla maaseudulla*. Helsinki: Nuorisotutkimusseura (Nuorisotutkimusseura, verkkojulkaisuja 117, Liike), pp. 23–40. Available at: https://www.nuorisotutkimusseura.fi/images/kuvat/verkkojulkaisut/reunamerkintoja_hylkysyrjasta.pdf.

Heleniak, T. (2020) 'Migration and mobility: more diverse, more urban', in Grunfelder, J., Norlén, G., Randall, L. and Sánchez Gassen, N. (eds) *State of the Nordic Region 2020*. Copenhagen: Nordic Council of Ministers (Nord, 2020:001), pp. 40–51. Available at: https://pub.norden.org/nord2020-001/nord2020-001.pdf.

Helminen, V., Nurmio, K., Rehunen, A., Ristimäki, M., Oinonen, K., Tiitu, M., Kotavaara, O., Antikainen, H. and Rusanen, J. (2014) *Kaupunki-maaseutu-alueluokitus. Paikkatietoihin perustuvan alueluokituksen muodostamisperiaatteet.* 25/2014. Suomen ympäristökeskus. Available at: https://helda.helsinki.fi/bitstream/handle/10138/135861/SYKEra_25_2014.pdf?sequence=1&isAllowed=y.

Helminen, V., Nurmio, K. and Vesanen, S. (2020) *Kaupunki-maaseutu-alueluokitus 2018. Paikkatietopohjaisen alueluokituksen päivitys.* Helsinki: Suomen Ympäristökeskus (Suomen ympäristökeskuksen raportteja, 21).

Horváth, D. I. (2008) 'The culture of migration of rural Romanian youth', *Journal of Ethnic and Migration Studies*. Routledge, *34*(5), pp. 771–786. doi: 10.1080/13691830802106036.

Jamieson, L. (2000) 'Migration, place and class: Youth in a rural area', *The Sociological Review*, *48*(2), pp. 203–223. doi:10.1111/1467-954X.00212.

Jänkälä, S. (2019) *Matkailun toimialaraportti*. Sarjajulkaisu 2019:3. Helsinki: Työ- ja elinkeinoministeriö. Available at: http://julkaisut.valtioneuvosto.fi/handle/10024/161292 (Accessed: March 19 2020).

Jankowiak, W. (2009) 'Well-being, cultural pathology, and personal rejuvenation in a Chinese city, 1981–2005', in Mathews, G. and Izquierdo, C. (eds) *Pursuits of happiness: well-being in anthropological perspective*. New York: Berghahn Books.

Jungsberg, L., Turunen, E., Heleniak, T., Wang, S., Ramage, J. and Roto, J. (2019) *Atlas of population, society and economy in the Arctic*. Stockholm: Nordregio (Nordregio Working Paper, 2019:3). Available at: http://urn.kb.se/resolve?urn=urn :nbn:se:norden:org:diva-5711 (Accessed: March 9 2020).

Karlsdóttir, A., Heleniak, T. and Kull, M. (2020) 'Births, children and young people', in Grunfelder, J., Norlén, G., Randall, L. and Sánchez Gassen, N. (eds) *State of the Nordic Region 2020*. Copenhagen: Nordic Council of Ministers (Nord, 2020:001), pp. 28–39. Available at: https://pub.norden.org/nord2020-001/nord2020-001.pdf.

Kiilakoski, T. (2016) *I am fire but my environment is the lighter. A study on locality, mobility, and youth engagement in the Barents region*. Helsinki: The Finnish Youth Research Society (The Finnish Youth Research Society Internet publications, 98). Available at: https://www.nuorisotutkimusseura.fi/images/julkaisuja/i_am_fire_ but_my_environment_is_the_lighter.pdf.

Kivijärvi, A. and Peltola, M. (2016) 'Johdanto. Sukupolvisten suhteiden liikuttamat lapset ja nuoret', in Kivijärvi, A. and Peltola, M. (eds) *Lapset ja nuoret muuttoliik-keessä. Nuorten elinolot -vuosikirja 2016*. Helsinki: Unigrafia (Nuorisotutki-musverkoston julkaisut), pp. 5–14. Available at: https://www.julkari.fi/bitstream/ handle/10024/131342/Nuorten%20elinolot%20-vuosikirja%202016_WEB. pdf?sequence=1.

Komu, T. (2019) 'Dreams of treasures and dreams of wilderness – engaging with the beyond-the-rational in extractive industries in northern Fennoscandia', *The Polar Journal*, 9(1), pp. 113–132. doi: 10.1080/2154896X.2019.1618556.

Kuntaliitto (2019a) *Kaupunkien ja kuntien lukumäärät ja väestötiedot*. www.kuntali-itto.fi. Available at: https://www.kuntaliitto.fi/tilastot-ja-julkaisut/kaupunkien-ja-kuntien-lukumaarat-ja-vaestotiedot.

Kuntaliitto (2019b) *Väestörakenne*. www.kuntaliitto.fi. Available at: https://www.kun-taliitto.fi/tilastot-ja-julkaisut/kuntakuvaajat/vaesto.

Lundgren, A. and Cuadrado, A. (2020) 'Wellbeing in the Nordic region', in Norlén, G., Randall, L. and Sánchez Gassen, N. (eds) *State of the Nordic Region 2020*. Copenhagen: Nordic Council of Ministers (Nord, 2020:001), pp. 130–141. Available at: https://pub.norden.org/nord2020-001/nord2020-001.pdf.

Malkki, L. (2012) *Kulttuuri, paikka ja muuttoliike*. Tampere: Vastapaino.

Massey, D. S., Arango, J., Hugo, G., Kouaouci, A., Pellegrino, A. and Taylor, J. E. (1993) 'Theories of international migration: A review and appraisal', *Population and Development Review*, 19(3), pp. 431–466. doi: 10.2307/2938462.

Mathews, G. (2009) 'Finding and keeping a purpose in life: Well-being and Ikigai in Japan and elsewhere', in Mathews, G. and Izquierdo, C. (eds) *Pursuits of happiness: well-being in anthropological perspective*. New York: Berghahn Books, pp. 167–186.

Mathews, G. and Izquierdo, C. (2009) 'Introduction: Anthropology, happiness, and well-being', in Mathews, G. and Izquierdo, C. (eds) *Pursuits of happiness: well-being in anthropological perspective*. New York: Berghahn Books, pp. 1–20.

Matkailun tutkimus- ja koulutusinstituutti (2013) *Matkailulla maakunta menestyy: Matkailun tulo- ja työllisyysvaikutukset 12 lappilaisessa kunnassa vuonna 2011*. Rovaniemi: Lapin yliopistopaino.

Metsähallitus (2019) *Kansallispuistojen, valtion retkeilyalueiden ja muiden virkistys-käytöllisesti merkittävimpien Metsähallituksen hallinnoimien suojelualueiden ja ret-keilykohteiden käyntimäärät vuonna 2019*, Metsähallitus. Available at: https://www. metsa.fi/documents/10739/3335805/kayntimaarat_2019.pdf/f5e5cfd2-ebad-4c4a-abee-19bd6f4a641d.

Moisio, J., Gissler, M., Haapakorva, P. and Myllyniemi, S. (2016) 'Lasten ja nuorten muuttoliike tilastoissa', in Kivijärvi, A. and Peltola, M. (eds) *Lapset ja nuoret muuttoliikkeessä. Nuorten elinolot -vuosikirja 2016*. Helsinki: Unigrafia (Nuorisotutkimusverkoston julkaisut), pp. 17–50. Available at: https://www.julkari.fi/bitstream/handle/10024/131342/Nuorten%20elinolot%20-vuosikirja%202016_WEB.pdf?sequence=1.

Myllyniemi, S. (2016) 'Tulevaisuudennäkymät', in Myllyniemi, S. (ed.) *Katse tulevaisuudessa. Nuorisobarometri 2016*. Helsinki: Grano Oy (Nuorisotutkimusverkoston julkaisut), pp. 19–34. Available at: https://tietoanuorista.fi/wp-content/uploads/2017/03/Nuorisobarometri_2016_WEB.pdf.

Nilsson, K. and Jokinen, J. K. (2020) 'Introduction', in Grunfelder, J., Norlén, G., Randall, L., and Sánchez Gassen, N. (eds) *State of the Nordic Region 2020*. Copenhagen: Nordic Council of Ministers (Nord, 2020:001), pp. 14–24. Available at: https://pub.norden.org/nord2020-001/nord2020-001.pdf.

Ollila, A. (2008) *Kerrottu tulevaisuus. Alueet ja nuoret, menestys ja marginaalit*. Rovaniemi: Lapin yliopistokustannus (Acta Universitatis Lapponiensis, 141). Available at: https://lauda.ulapland.fi/handle/10024/61759.

Palen, R. and Lien, H. (2018) *Young people on the labour market in 2016*. Eurostat. Available at: https://ec.europa.eu/eurostat/documents/2995521/8768233/3-27032018-AP-EN.pdf/3a8861db-939c-4790-a3bc-8837bbbac15c.

Pedersen, H. D. and Gram, M. (2018) '"The brainy ones are leaving": The subtlety of (un)cool places through the eyes of rural youth', *Journal of Youth Studies*, 21(5), pp. 620–635. doi: 10.1080/13676261.2017.1406071.

Penttinen, P. (2016) 'Viettäisinkö elämäni periferiassa? Nuorten muuttoalttius Itä-Suomessa', in Kivijärvi, A. and Peltola, M. (eds) *Lapset ja nuoret muuttoliikkeessä. Nuorten elinolot – vuosikirja 2016*. Helsinki: Unigrafia (Nuorisotutkimusverkoston julkaisut), pp. 149–165. Available at: https://www.julkari.fi/bitstream/handle/10024/131342/Nuorten%20elinolot%20-vuosikirja%202016_WEB.pdf?sequence=1.

Pitkänen, K., Sireni, M., Rannikko, P., Tuulentie, S. and Hiltunen, M. J. (2017) 'Temporary mobilities regenerating rural places. Case studies from Northern and Eastern Finland', *Journal of Rural and Community Development*, 12(2–3), pp. 93–113.

Ristikari, T., Törmäkangas, L., Aino, L., Haapakorva, P., Kiilakoski, T., Merikukka, M., Hautakoski, A., Pekkarinen, E. and Gissler, M. (2016) *Suomi nuorten kasvuympäristönä. 25 vuoden seuranta vuonna 1987 Suomessa syntyneistä nuorista aikuisista*. 9/2016. Tampere: Juvenes Print - Suomen yliopistopaino Oy.

Robben, A. C. G. M. and Sluka, J. A. (eds) (2012) *Ethnographic fieldwork: An anthropological reader*. West Sussex: John Wiley & Sons.

Robertson, S., Harris, A. and Baldassar, L. (2018) 'Mobile transitions: a conceptual framework for researching a generation on the move', *Journal of Youth Studies*, 21(2), pp. 203–217. doi: 10.1080/13676261.2017.1362101.

Rönnlund, M. (2019) '"I love this place, but I won't stay": Identification with place and imagined spatial futures among youth living in rural areas in Sweden', *Young - Nordic Journal of Youth Research*, pp. 1–15. doi: 10.1177/1103308818823818.

Sapolsky, R. M. (2004) *Why zebras don't get ulcers*. New York: Henry Holt and Company, LLC. Available at: https://www.audible.co.uk/pd/Why-Zebras-Dont-Get-Ulcers-Audiobook/B00ATSJBJO.

Sardan, J.-P. O. de (2015) *Epistemology, fieldwork, and anthropology*. New York: Palgrave Macmillan.

Seyfrit, C. L., Bjarnason, T. and Olafsson, K. (2010) 'Migration intentions of rural youth in Iceland: Can a large-scale development project stem the tide of out-migration?', *Society & Natural Resources, 23*(12), pp. 1201–1215. doi: 10.1080/08941920903278152.

Similä, J. and Jokinen, M. (2018) 'Governing conflicts between mining and tourism in the Arctic', *Arctic Review, 9*, pp. 148–173. doi: 10.23865/arctic.v9.1068.

Sireni, M., Halonen, M., Hannonen, O., Hirvonen, T., Jolkkonen, A., Kahila, P., Kattilakoski, M., Kuhmonen, H.-M., Kurvinen, A., Lemponen, V., Rautiainen, S., Saukkonen, P. and Åström, C. (2017) *Maaseutukatsaus 2017. Maa- ja metsätalous-ministeriön julkaisuja 7/2017.* Helsinki: Maa- ja metsätalousministeriö. Available at: http://urn.fi/URN:ISBN:978-952-453-958-6.

Stockdale, A. and Haartsen, T. (2018) 'Editorial introduction: Putting rural stayers in the spotlight', *Population, Space and Place, 24*(4), pp. 1–8. doi: 10.1002/psp.2124.

Taylor, C. (1989) *Sources of the self: The making of the modern identity.* Cambridge: Harvard University Press.

Thin, N. (2009) 'Why anthropology can ill afford to ignore well-being', in Mathews, G. and Izquierdo, C. (eds) *Pursuits of happiness: Well-being in anthropological perspective.* New York: Berghahn Books, pp. 23–44.

Thin, N. (2016) 'Home and away: Place appreciation and purposeful relocation in later life', *Anthropology in Action. Berghahn Journals, 23*(3), pp. 6–16. doi: 10.3167/aia.2016.230302.

Tuhkunen, A. (2007) *Between location and a sense of place: Observations regarding young people's migration alacrity in Northern Europe.* Tampere: Tampereen yliopistopaino Oy (Acta Universitatis Tamperensis, 1207).

Tuuva-Hongisto, S., Pöysä, V. and Armila, P. (2016) *Syrjäkylien nuoret - unohdetut kuntalaiset?* Keuruu: Otavan kirjapaino Oy (Tutkimusjulkaisu-sarjan julkaisu, 99). Available at: https://kaks.fi/wp-content/uploads/2016/10/Syrj%C3%A4kylien-nuoret-unohdetut-kuntalaiset.pdf.

Valtion nuorisoneuvosto (2016) *I am prepared to move to a different town or city to get a job, tietoanuorista.fi.* Available at: https://nuorisobarometri.tietoanuorista.fi/olen-valmis-vaihtamaan-asuinkuntaa-tyopaikan-saamiseksi (Accessed: March 25 2020).

Valtion nuorisoneuvosto (2019) *Nuorten määrä, tietoanuorista.fi.* Available at: https://indikaattorit.tietoanuorista.fi/taustaindikaattorit/nuorten-maara-2 (Accessed: March 25 2020).

Wrede-Jäntti, M. (2020) *Nabo – nuorten kokemuksia sosiaalisesta osallisuudesta Suomessa.* Copenhagen: Pohjoismaiden Ministerineuvosto (TemaNord, 2019:556). Available at: http://norden.diva-portal.org/smash/get/diva2:1393114/FULLTEXT01.pdf (Accessed: Accessed: March 9 2020).

Ministry of Education of Finland (2017) *Finnish Youth Act.* Available at: https://minedu.fi/documents/1410845/4276311/Youth+Act+2017/c9416321-15d7-4a32-b29a-314ce961bf06/Youth+Act+2017.pdf (Accessed: November 23 2020).

3 Leaving or staying?

Youth agency and the liveability of industrial towns in the Russian Arctic

Alla Bolotova

All that surrounds us is the mining enterprise.
[Всё, что нас окружает – это ГОК].

(female, 32, Revda)

Introduction

All my friends have a similar opinion: we should get out of here. However, the opportunity to leave is a different thing. The desire to leave is one thing, and the possibility to leave is another thing.

(male, 29, Kirovsk)

These words were said to me by a young miner in a single-industry town in the Russian Arctic. In northern industrial communities, most young people ponder how to leave their hometowns (Vakhtin and Dudeck 2020). The idea that youth must migrate to big cities for a successful life is not confined to the Arctic, however. In many areas around the world, the massive outmigration of young people from peripheral communities has become commonplace and normalized (e.g., Corbett 2007; Leibert 2016; Farrugia 2014). Across the Arctic, the large-scale outmigration of youth is a reality not just in Russia, but across all circumpolar countries. In the industrialized Russian Arctic, the expectation that young people will move away after finishing school is widespread even in towns with economically profitable and stable working enterprises. This outmigration of local youth is widely supported by all generations, it is simply in the air: parents, relatives, schoolteachers, older friends all repeat the narrative.[1] This widespread orientation to outmigration is connected to the history of Soviet industrial towns in the North that were settled and populated by work migrants from all over the Soviet Union. Consequently, a "culture of migration" endures in the public discourse that urges young people to move away (Ali 2007; Horváth 2008, see Komu and Adams, this volume).

There are two main reasons usually cited with regard to youth outmigration: the lack of job opportunities and limited possibilities for education (Stockdale 2002). These factors are also important in northern mono-industrial towns, though the main problem in these towns is not a shortage of jobs, but, rather, a small variety of available choices. Nevertheless, as was

DOI: 10.4324/9781003110019-5

expressed by the mining worker cited at the beginning of this chapter, not all young people who want to leave have an opportunity to do so, and even some of those who have left might return to their hometowns later, after living elsewhere for a period of time. Every young "stayer" has wrestled with the decision to stay or to leave their northern hometown multiple times over their life course and are adept at dealing with the complexities of leaving/staying within the local "culture of migration". While the amount of research on outmigration and migrants is rather large, the staying process and agency of "stayers" are still understudied (Schewel 2020). In the migration discourse, staying is often portrayed negatively, being associated with passivity, traditional values and expectations (Stockdale and Haartsen 2018).

The aim of this chapter is to analyze the migratory decisions, life choices and agency of stayers and returners, exploring them in the contexts of specific circumstances, structural constraints and opportunities in single-industry towns in the Russian Arctic. The chapter asks the following question: what are the historical roots and current reasons for the "orientation toward leaving" common among the young generation in the urbanized Russian Arctic? What kind of choices do stayers make when they decide to stay or to return to their hometowns, and how do processes at the town's main enterprises (town-forming companies, see Adams et al., this volume) influence these decisions? The contribution focuses, in particular, on the agency of stayers and explores staying as a process, showing it through a series of ethnographic vignettes on the following issues: gendered work in mining; self-employed entrepreneurship; and urban activism.

This chapter is based on two months of ethnographic fieldwork that I conducted for the WOLLIE research project in several mono-industrial towns in the Murmansk Region,[2] as well as on my previous long-term fieldworks in the region.[3] These towns are all connected to stable mining industries and situated next to deposits of different mineral resources: iron, apatite, and rare earth minerals. The population in these towns varies from 7,900 people in Revda to 26,206 people in Kirovsk (Rosstat 2020), and all the cities studied have experienced a high level of outmigration in recent decades (Monogoroda Arkticheskoi zony RF 2016). The corpus of interviews (n = 85) consists of unstructured in-depth interviews with young people aged between 18 and 35,[4] supplemented by interviews with representatives of local administrations, the town-forming companies and other important community institutions. The interviews were recorded and transcribed. In order to protect the privacy of interviewees I changed the names of most of the participants, with the exception of two urban activists who allowed me to mention their names in publications (Nastya and Valera, who appear in the section "Activism and liveability in single-industry towns").

Peripheralization and agency of youth in declining communities

Massive outmigration of youth from the shrinking industrial towns in the Russian Arctic is linked with the global phenomenon of peripheralization, which is especially strong in circumpolar contexts (Heleniak and Bogoyavlensky 2015; Heleniak 1999; Bjarnason and Thorlindsson 2006; Rasmussen 2007; Martin 2009). Uneven spatial development is one of the most urgent contemporary global challenges (Smith 1984; Brenner 2004, 2019). Due to peripheralization and economic polarization, the number of spatially disadvantaged localities continues to increase (Kühn 2015; Haase et al. 2014; Martinez-Fernandez et al. 2012; Weaver et al. 2016). Declining industrial communities exist in various places in Russia and former Soviet republics (e.g., see Pelkmans 2013), as well as in other countries around the world (e.g., Lansbury and Breakspear 1995; Robertson 2006). Declining cities not only lose their population, but also experience a decrease in investments and state support and service cutbacks, which can cause a cycle of decay. The social development of cities in decline is characterized by accelerated ageing and the loss of the working-age and better-educated population; the so-called "brain drain" (Haase et al. 2014). Public infrastructure, such as schools, health services, or public transport, is often reduced so as to cut the costs of maintenance. Moreover, urban shrinkage causes material consequences (e.g., abandonment and vacancy, deteriorated buildings, spatial marginalization) that diminish community liveability and foster a sense of resignation and powerlessness (Ringel 2018; Mah 2012).

Declining peripheral cities have a rather negative image in public discourse (Kinder 2016; Béal et al. 2017). In the media, shrinking cities are usually portrayed as "problematic" sites with a high level of deprivation that, in extreme cases, leads to a stigma of "dying cities": places without resources inhabited by deprived people or the "losers of society" who cannot escape (Bernt and Rink 2010). This stigmatization influences the self-perception and self-respect of these localities.[5] In short, material losses in such communities are almost always accompanied by a loss of dignity and self-worth.

As result, in declining towns, many residents of different ages view the possibility to move out positively. As noted above, employment deficits and lack of education are the most powerful driving forces pulling youth away. However, it is important to look beyond these structural constraints and address questions of youth agency in making migratory choices. Kerilyn Schewel emphasizes a mobility bias in research, arguing that migration theories neglect the countervailing structural and personal forces that restrict or resist "drivers" of migration and lead to different immobility outcomes (Schewel, 2020). There is an increasing trend in migration research to focus on non-migrants, or these who stay put (Gray 2011; Fernandez-Carro and Evandrou 2014; Hjalm 2014; Mata-Codesal 2015; Preece 2018). These researchers are aiming to rethink immobility and see staying as an active process, in which stayers are considered not as passive observers of their fates, but as active participants (Stockdale and Haartsen 2018).

The process of staying is as nuanced and diverse as the process of migrating and the decision to stay is made multiple times over an individual's life course (Hjalm 2014; Stockdale and Haartsen 2018). This decision is made under a particular combination of structural influences on the agency of actors who may respond to the same forces in different ways. The processes, experiences and perceptions of staying depend largely on the degree of (in) voluntariness. Immobility can be experienced either as a burden or as an achievement (Mata-Codesal 2015). Therefore, researchers make a distinction between immobility as a nuanced choice ("stillness") versus a product of constraints ("stuckness") (Cresswell 2012). An involuntary immobile is a person who would like to move but lacks the ability and means to do so, and therefore feels stuck or left behind (Carling 2002; Stockdale and Haartsen 2018). Desired immobility is a consequence of conscious decisions of individuals to attach themselves to places. This choice is based on a complex combination of social, emotional and material factors (Hjalm 2014; Mata-Codesal 2015).[6] However, agency still exists in involuntary immobility. Saba Mahmood defines agency as a capacity for action that is created and enabled by relations of subordination (Mahmood 2001). Here I explore the agency of stayers in the multiple ways in which young people make choices about life, choosing different paths and shaping their life course trajectories in particular contexts. Institutional regulations, structural opportunities and constraints, individual (and family) agency, and emotions are related and intertwined, influencing the way young adults take control of their lives (Evans 2002, 2007).

Despite a large number of involuntary immobile young people in declining peripheral communities, there are also young people who make a conscious choice to stay and who have become activists in the urban life of their cities. These active young citizens are involved in various projects intended to create new initiatives to improve the city's liveability. In general, in each marginalized peripheral city, there are some people actively struggling with abandonment, the loss of local services and facilities and trying to create various local initiatives to make the cities more viable (Corcoran 2002). Such bottom-up initiatives can create jobs, albeit usually in small numbers, improve physical space, stabilize the sense of community, and bring spirit and hope to the communities.

In studies of urban liveability, emphasis is often put on the economic aspects, examining urban life within the narrow framework of the city's physical dimensions, particularly infrastructure, urban services and economic growth, but paying less attention to local cultures, community bonds and the associational life of communities (Ho and Douglass 2008; Kong 2009). Studying urban initiatives of youth in declining northern cities, I follow researchers who argue that liveability should not be reduced solely to the material or economic wealth of cities (Ho and Douglass 2008) and that more attention should be paid to the social relations, community life and human agency needed to make a city liveable for different social groups, including youth.

Soviet single-industry towns in the Arctic and their post-Soviet transformations

How did a local "culture of migration" (see Komu and Adams, this volume) form in northern industrial towns? Research shows that individual migration decisions are always deeply rooted in local contexts and practices. To understand the historical roots of the "orientation toward leaving" common for young people, I briefly explore how the towns were built from scratch during the Soviet period and how they have changed over time.

Before the Soviet period, the Russian Arctic was scarcely populated. Today, it boasts the most industrialized and urbanized polar territory in the world, containing 72% of the circumpolar Arctic population (Rasmussen 2011). During Soviet industrialization, numerous new mining towns were built from the ground up near rich deposits of valuable mineral resources. In the Murmansk Region, where this study was conducted, the first new industrial cities were founded in the 1930s and, to a large extent, they were built by forced workers: Gulag prisoners and exiled peasants (Shashkov 2004; Bolotova and Stammler 2010; Bolotova 2014). After the death of Stalin and the dissolution of the Gulag system, even more new cities were built; however, they were populated by voluntary migrants from all over the Soviet Union, attracted to the North by material benefits.

During the second (voluntary) wave of industrialization, northern single-industry towns became prosperous multi-ethnic communities of work migrants in which the town-forming company was the "owner" of the city and structured practically all community institutions, controlling not only industrial production but all other spheres of life. Even housing, health care, sport, and culture functioned as provisions of the enterprise. In other words, work constituted the whole life of individuals and also took the central social position in urban life of the communities.

Following the dissolution of the Soviet Union, all northern mining communities went through drastic changes caused by the radical social and economic transformations occurring across Russia. At this time, the state subsidies to northern single-industry towns were reduced significantly. All communities faced the radical modernization of their local industrial enterprises and the large-scale restructuring of their economic and social spheres (Pilyasov 2013, p. 3). In the Murmansk Region, most town-forming companies were privatized and came under the control of large corporations. In some cases, the owners of the main enterprise have changed on several occasions. This radical change from prosperity and stability to post-Soviet uncertainty and crisis in many northern industrial towns caused large-scale outmigration and downsizing, including the case study cities in the Murmansk Region. Between 1989 and 2019, for example, the population of Kirovsk decreased by 39.8%; Revda by 42.5%; and Kovdor by 46%.[7]

Today the town-forming mining companies are mostly managed from the outside: the top managers who make decisions usually reside in other places and visit these towns only occasionally; consequently, they do not have personal connections to the localities. In contrast to the initial Soviet-era development of company towns in which social and cultural life was integrated with economic development, the post-Soviet enterprises have become less active in the local community life. The large companies that succeeded the former state enterprises routinely try to minimize their social burdens and avoid additional expenses. There have been several stages of restructuring and neoliberal reorganization of local enterprises aimed at increasing productivity and reducing operational costs through the implementation of outsourcing and subcontracting strategies. This has led to a decrease in social benefits and lower salaries for workers (Suutarinen, 2015). The restructuring has also entailed a radical reduction in the numbers of workers since now the efficient operation of mines needs less workforce than before, because of the automation of various processes and changes in mining technologies. For example, in Kirovsk the Apatit company fired more than 2,000 employees in 2013 and 3,000 more in 2014. The fired workers were either moved out of the company into subcontracting companies, retired or migrated to other places. As a result, the total number of the company's employees dropped from 11,600 in 2012 to 7,100 by the beginning of 2015 (Didyk 2015, p. 4). This continuous restructuring has led to further degradation in the employment conditions of mining workers. Workforce reduction tendencies influence most the weakest groups in the local communities. Young people are especially vulnerable to the changes, because of their lower qualifications at the start of their working life. Working pensioners and young women are also especially in danger under these conditions of ongoing neoliberal "optimizations and restructuring", due to the age and pregnancy discrimination practiced by employers.

Despite the successful internationalization of the mining enterprises, these single-industry towns built during the Soviet period are still very much rooted in the Soviet past. This rooting continues to shape the lives of contemporary youth and influences their migration decisions and life choices.

Staying or leaving?

Young people living in northern industrial cities routinely face the question of whether to stay or leave many times throughout their lives, especially at major turning points of their biographies, e.g., finishing schools, getting married, becoming parents. At these life points, people frequently and ubiquitously discuss ideas and plans to move out—from casual conversations in the streets and bars to more serious conversations with friends and family. Those who have no plan to leave at least have to wrestle with the idea of moving elsewhere in the future. This is exacerbated because many of the young people in northern industrial cities grow up in highly mobile families of former work migrants who regularly travel outside their home regions since people working in the North have rather long vacations.[8] As a result, many northern residents develop attachment and connections to both their northern homes and their regions of origin (Liarskaia et al. 2020; Bolotova et al. 2017). In other words, most young people who grow up in the North are used to travel and are mobile from early childhood. Furthermore, it is common for northern youth to encounter many people from older cohorts who have left but return to visit. Consequently, their aspirations for the future are formed in an environment in which outmigration and mobility constitute important parts of their life worlds. A high level of spatial mobility from an early age and exposure to a "culture of migration" prepares young people in the northern towns for outmigration.

However, outmigration is rarely an individual choice. Rather, it is a shared decision, formed in the local social worlds of young people, where moving out is largely supported by older generations. In many cases, the future out-migration of a young person is planned and prepared in advance by older members of the family. Such a paternalistic approach to youth is relatively common in Russia, and this decision is not primarily an economic choice for securing the parents' future in old age, but more a result of strong family ties. Parents of school-aged children typically already try to buy apartments for kids in their future places of education, planning their relocation in advance.

> There is a massive purchase of apartments for children here. Everyone wants to go somewhere, mostly to St. Petersburg. [...] I have many acquain-tances, whose children are only ten years old, but they already bought an apartment in St. Petersburg. One of my friends has two kids, the first is six years old, the second just turned one, and she already bought an apart-ment in St. Petersburg, and they are already saving money for the second apartment.
>
> (female, 35, Kovdor)

After finishing school, young people in northern industrial cities are faced by a lack of post-school educational opportunities in their hometowns. The existing options of post-secondary education in northern towns provide a very narrow choice of professions, the majority of which are related to min-ing. In fact, the orientation to mining starts even in primary school educa-tion, because many town-forming mining companies actively support mining-oriented education in the towns where they operate. For example, in Kirovsk and Apatity the town-forming company Apatit (belongs to the PhosAgro chemical holding company) sponsors specialized education and training through their so-called PhosAgro classes, which provide privileged educational conditions for children who choose to study in these classes. Employing a variety of benefits, the company stimulates children to study mining-related subjects, such as chemistry, math, physics and computer sci-ence, and pushes them to enter technical universities to obtain professions that are in demand at the company. Such narrow specialization is not inter-esting for many of today's young people, so youth with interests other than mining often seek opportunities in other places.

The spectrum of jobs available in northern cities is also very limited and often restricted to mining. Young women have especially limited job chances since mining in Russia is traditionally a male-dominated industry with a mas-culine occupational culture. Job opportunities for young women are available mostly in traditionally female-dominated sectors, such as different types of services, healthcare, administration or education. Gender differences in the labour market significantly influence the life trajectories of young women who leave in larger numbers than men. "There are no big prospects in our city. All the girls either go to work in shops or kindergartens. All other spheres are occupied" (female, 16, Revda).

Despite the general importance of the economic drivers of outmigration, there are also other factors beyond economic rationalities that influence young people's life choices. Youth living in northern industrial towns often complain about a lack of recreational opportunities in hometowns, boredom,[9] and the Soviet appearance of urban space in their localities: "Of course all my friends are going to leave! It is too boring here. And there are no good cafes and museums as in St. Petersburg" (female, 17, Kovdor). In some localities, the material decay of the infrastructure and the urban environment contributes to the negative image of their home localities in the eyes of young people. They are lured by big cities and are sensitive to the stigma of declining cities common in the public discourse. "All that we want is to go to St. Petersburg. We want to try something new, and if we do not succeed, we will return here" (female, 15, Kirovsk). In short, the future is largely perceived to be elsewhere, not in the hometown.

The majority of young northerners have a definite answer on the question of whether they will leave or stay after they graduate from schools: "we should get out of here". In addition to the lack of jobs and educational opportunities, other factors also have a significant influence on young people, e.g., the attraction of big cities, the negative images of declining cities, a "culture of migration" and their parents' opinions. This begs the question of why some young people, who are also embedded in the northern "culture of migration", do *not* migrate in this age of pervasive migration or put off their relocation to the distant future? Staying also requires agency and this is a complicated decision made many times in the life course of individuals. Below I explore empirically the varying lived experiences of immobility, considering a variety of situations and paths of stayers, from involuntary immobility (Carling 2002), or "stuckness" (Cresswell 2012), to conscious decisions that the North will be their place of residence and local activism.

Stuckness: *"Nothing can save this place!"*

Involuntary immobility is the situation in which an individual wants to move, but lacks the ability and/or means to accomplish this relocation. In marginalized and disadvantaged communities, many young people express a desire to leave; however, many of them cannot organize the move and are kept put by various structural constraints. The story of Sergej from Revda (21) is a good example of a person who got "stuck" in his hometown. He works at the local prison as a dog handler. Sergej does not like his job, which he describes as emotionally difficult and physically hard due to the long shifts. For a long time, Sergej has been dreaming of moving out of Revda:

> I do not like it here because there is nothing to do in Revda. There are two things: nothing to do in Revda and I am not happy with the job. If

Revda would suit me, then I would work there, okay. But these two factors! And in the end, it is very sad here, ideally, I want to leave [...].

At the same time, Sergej is very attracted by the local natural environments, often spending time in the nearest mountains, but this does not reconcile him with his situation:

Nature is very beautiful here, I like it a lot. I do not cease to admire the landscapes of Revda and the Murmansk Region in general. There are very few such settlements where you could leave your house and see two mountains right in front of you. This, of course, is beautiful, and the air here is excellent, but it only makes up for a bit that there is nothing to do at all.

Most of Sergej's close friends have already left, either to Murmansk (the region's major city) or to St. Petersburg: "Literally all my peers are gone now, only a few stayed here. You enter into such a situation that you make good friends with the current 11th-grade school graduates, but very soon they will also leave." Sergej tries to plan his future move; however, he does not have any professional education because he started to work immediately after he served his mandatory term in the army. This lack of education limits his work opportunities:

I do not want to work at our mining enterprise. There you can work either on the surface or in the underground mine, but in the underground, you can die. Once every couple of years, an incident occurs that someone was crushed there or the stone plate fell, or gas exploded, so it is really very dangerous there. And if you work on the surface, the salary is very low. The compromise is to work at the grocery store, where salaries are a bit higher. And that's all, I don't know where else I could go here without education.

In addition, Sergej's parents would find it difficult to support his further education and relocation, and his salary is insufficient to cover anything other than basic living costs. He feels "left behind" and stuck in his hometown because of the high level of youth outmigration. Despite all these difficulties, Sergej still hopes to leave one day. He is thinking about how to get a professional education in the future. He fosters active connections with his old friends who have already moved, exploring different options and opportunities for relocation, and thereby trying to increase his chances to migrate.

Working in mining as a life choice

An orientation toward mining is very common in mining towns, especially among the male population. Many young people simply follow the familiar life strategies of their parents; while others use the numerous mining-related educational and career opportunities provided by town-forming companies. Some people might still consider outmigration as a possible choice for the future but chose to stay for their current life stage. "Frankly speaking, I would actually go to St. Petersburg, if I had something there: some kind of 'airbag', like an apartment, or for example, good work with a high salary. To go there having nothing is not for me. Here I have a good job, a salary, an apartment. To throw away all of this and leave for somewhere else—no, of course not" (mining worker, 24, Kirovsk/Apatity). Working in mining and staying in hometowns provides young people with a kind of stability and standard of living that they would be unable to maintain in big cities.

However, this kind of immobility can also be viewed as temporary. For example, changes in the town-forming companies might significantly influence the life plans of young people and push them towards outmigration. In this case, moving is a direct reaction to reorganization processes at the company: "The policy of the [town-forming] company has become quite inadequate. [...] Highly qualified people are leaving, I don't know where, they are just leaving. Some of them go to work at subcontracting companies, others just leave the city" (male, mining worker, 29, Kirovsk). This growing precarity, instability and insecurity at the workplace influences young people when they make decisions to leave or to stay.

Women's paths in mining cities

Mining in Russia is still a very male-dominated industry, which causes particular difficulties for young women seeking employment in cities dominated by extractive industries. Many young women experience gender discrimination, such as when hiring managers in mining companies give preference to male candidates, even if the female applicants have higher qualifications. The emphasis on mining at schools also stimulates young girls with other interests to relocate to receive a broader education and subsequently more possibilities.

> I plan to go to St. Petersburg to study there and I will stay there if it will be possible thereafter. I think there are not many opportunities in our town. All opportunities are connected to working in mining, associated with mining professions. I do not know yet what I want to do, but I definitely do not want to work in the mine.
>
> (female, 15, Kirovsk)

Because life circumstances change, it is common for girls who go away after school to later return to their hometowns. Some come back to take care of sick parents, others because of a relationship with somebody from the hometown, or because it proved too difficult to survive in a big city. Upon returning, they often they face the problem of finding a job in the mining city. For example, Maria (35) from Kovdor went to study law in the regional center Murmansk, but later got married and returned to Kovdor to live with her husband. She finished her study through remote education, got a lawyer's diploma and gave birth to her first child. After maternity leave, she started to search for a job.

> And then I went through all the troubles and ordeals that were ever possible in a small mining town for a female lawyer. It was impossible. I experienced so many problems, I cried so much! You should understand, it was simply unreal. But I was fighting and knocking at all doors.

The first job she was able to get after she received her law degree was the job as a cleaner at the local court. She was told that the cleaning would not take much time and after she finished these main tasks, she could help the judge and other court personnel with their duties to get some practical experience in jurisprudence. Maria agreed and this was indeed her first practical step in becoming a professional lawyer. Later, she succeeded in getting several other jobs as a lawyer in local companies; however, she was also demoted on several occasions:

> I really liked the work of a lawyer, and I was very good in it. At my last workplace, everything was wonderful with the job, and I was happy working there, also the salary was OK. But then, at some point our main mining enterprise began to dictate conditions again, making a series of workforce reductions and reorganizations. This time they cut half the engineers and other qualified personnel, so again, I was laid off and lost the job. All in all, I had just four years of normal calm work at one enterprise and two years at another enterprise. Laid off here, laid off there, in the end, it put me at a low ebb. For me, it was a real blow.

This example demonstrates the particular vulnerability of women in mining cities with limited opportunities for other jobs than mining.

"Forced" entrepreneurship

In the conditions of growing precarity caused by short-term contracting and job reductions at town-forming mining companies, some young people have switched to self-employment, opening up new small businesses. In this way, declining job opportunities and limited alternatives in hometowns push young people toward entrepreneurship and self-employment. Many of them have no prior experience of entrepreneurship and had never dreamt about owning a business. Yet they were "forced" to try a new strategy by a combination of life circumstances and structural constraints (Oakley 2014). Below I consider several examples of how young entrepreneurs exercise their agency and how they make choices in the conditions of single-industry towns.

At one point the lawyer Maria from Kovdor, whose job search was recounted in the previous section, decided that she cannot continue searching for wage work in her city:

Once I told myself: "Damn, that's enough, I can't stand it all anymore. It's humiliating for me to go looking for a job here". At that time, I already understood that I am a pretty good specialist, I already had a lot of experience in the legal sphere. But I knew I had to change something. Then I began to think about what to change, what would I do if I worked for myself. In one month, I decided that I would completely change professions. I planned to open a business, but I did not want to get involved in jurisprudence. I just started looking for what I would be very much interested in doing. And I was always interested in girls' things.

Maria decided to open up a small-scale spa and beauty salon, where she could carry oit various beauty treatments for women. Her main idea was "to create a space for women where they could rest". Maria went to Moscow to get a professional education in cosmetics and, with the financial help of her husband, succeeded in starting up a spa. For several years, she combined working at her spa with a part-time salary job as a lawyer at one of the subcontracting mining companies. Later, she decided to quit the law job in order to focus entirely on her spa. The spa became relatively successful, even though Maria thinks there is no potential for further growth in Kovdor because in a mining city not many women are interested in this kind of treatment and relaxation.

Another case of "forced" entrepreneurship can be observed in the example of the married couple Sasha and Lena (both 29) from Kirovsk. They grew up in workers' families in which the men were working in the mine while their wives worked at various service jobs, such as cleaning, house painting and mail delivery. Following the dissolution of the Soviet Union, Sasha's mother started her own business and became quite successful. Now she owns a small sewing company producing jackets. Sasha and Lena met when they were studying pedagogy at the university in Apatity. Later, despite having a higher pedagogical education, Sasha worked at a wide variety of jobs. This included a post with the town-forming company Apatit, which he later lost due to cuts after a company reorganization. Sasha then decided to start his own business, though his wife did not support this decision:

She had an idea that I have to work at the mine. You know, her parents were there: her father worked as a shaftman, her mother was cleaning and working with wood there. In other words, she did not believe at all in entrepreneurship. She was crying. But I said, no, I want to try and will get it working, everything will be fine.

According to Sasha, his main example was his mother, who became an entrepreneur in the difficult 1990s while his father continued working in the mine and earned much less. Sasha had always liked his mother's lifestyle and wanted to follow her. After he got cut from the Apatit company, he registered at the job centre as a job seeker. Because he was unemployed, Sasha was able to get a state subsidy to open up a new private business. After two years of operating a fairly successful photo-printing company, Sasha received an invitation to return to the Apatit mining company. At this point he decided to close his small business and accept the offer because he was missing the stability of salaried work. His wife Lena, who at that time was on the maternity leave, decided to continue his photo-printing business and registered the individual entrepreneurship in her name. By that time, she had become convinced that being an entrepreneur had advantages compared with work in mining. Subsequently, Sasha quit the mining job and now the couple owns two small private businesses: one printing photos and one organizing events and activities for children. They are thinking about moving out of Kirovsk in the distant future, but are afraid to lose the social connections, support and networks that they have developed in Kirovsk. Consequently, they have not yet taken any concrete steps toward moving.

These ethnographic vignettes reveal that young adults living in mining cities choose self-employment and entrepreneurship not as a first choice, but as a response to the growing insecurity of wage work at the mining enterprises. While many young residents of mining towns chose outmigration as an alternative to labour market insecurity, some of those who stay become self-employed through what can be termed "forced" entrepreneurship. This kind of shift requires a great degree of agency from a young person because of the numerous structural constraints and the scarcity of opportunities and state support available for young entrepreneurs in Russia.

Activism and liveability in single-industry towns

How do young people who chose to stay in their hometowns participate in the development of the urban environment and contribute to the improvement of the liveability of northern cities? In general, urban civic activism is not particularly common in peripheral towns in Russia. Furthermore, inhabitants of single-industry towns are often portrayed as especially passive and lacking in civic engagement, expecting the state and the town-forming companies to solve all local problems. Below I explore youth civic activism using the example of one of my case study localities to focus on how young active "stayers" try to make their cities more liveable.

Kirovsk is a very popular destination for skiing tourism as it is situated next to the Khibiny Mountains. During the last decade, Apatit, the company which operates the town-forming mining operations, made large investments in the development of a ski resort in Kirovsk. Still, local youth often continue to complain about a lack of leisure opportunities in the city as the new ski facilities are mostly oriented towards tourists and visitors. Recently, loosely organized groups of young urban activists have become more visible in the public sphere of Kirovsk. They have initiated new projects to contribute to the development of the urban environment of Kirovsk according to their own needs. These initiatives are mostly apolitical in an institutional or state-centred sense, aimed at small improvements in living conditions at the local level, but they provide important alternatives to the mining industry and outdated Soviet-style activities. They also bring more diversity to the urban environment and public space. Some young people create new tourism and leisure opportunities oriented to both local residents and tourists. Others organize citizen groups and bottom-up events that improve the liveability of the city, such as volunteer networks to help elderly or poor people, informal civic groups to promote recycling, artistic groups to initiate music festivals, and sports competitions and adventure races. Below I consider two of these civic bottom-up initiatives in particular, namely the organization of a recycling initiative and the creation of a new open space for youth. In my analysis, I focus on the intentions and life choices of the individuals who initiated the projects rather than on the actual development of the initiatives, in order to demonstrate how a young person who grew up in a mining city may become an urban activist.

Recycling

Environmental issues are among the most topical for youth across the globe, including the Russian Arctic. Young residents of Kirovsk are significantly more environmentally concerned than older generations, though interaction with the environment during leisure time is very important for the majority of northern residents (Bolotova 2012). Two major factors contribute to the rising popularity of "green values" among the northern youth: the active use of the internet, which increases their awareness of global environmental concerns; and the high level of spatial mobility. Currently, one of the most pressing environmental issues in Russia is the problem of waste management. The current system of waste management is mainly oriented to landfill disposal of waste, with a low level of waste processing. There is almost no recycling infrastructure across Russia and many of the existing landfills are reaching their capacity limits. Young people concerned about environmental problems are increasingly dissatisfied with this situation. In different places in Russia, young environmental activists have organized grassroots groups to popularize recycling and selective collection of waste, as well as to pressure the authorities to establish recycling infrastructure. Recently, one such group was formed in Kirovsk.

The group was initiated by Nastya (28), who is a self-employed entrepreneur producing handmade polymer clay jewelry and accessories. Nastya was born in Kirovsk and lived there for most of her life, only travelling beyond the city for summer vacations. She never wanted to relocate from her hometown because she does not like big cities and hot weather and feels very comfortable in the North. She studied economics at the university in Apatity and then received some work experience as an accountant at the city administration. Then she shortly worked at the town-forming company Apatit, but quit because she did not like the stressful work conditions there. She established a small business developing her hobby and became rather successful, selling her handmade jewelry over the internet. About five years ago, Nastya became very environmentally concerned when she started to read a lot of information on the internet about environmental problems:

> When I became interested in environmental issues, recycling, and the current condition of the world, I first became very depressed. I realized that while there are maybe enough resources for me, already my children will not have enough, and my great-grandchildren will live on a landfill.

After some time, she decided to do everything that she could in her private life to reduce the amount of waste she created. She discovered several small companies in the Kirovsk-Apatity agglomeration that were interested in obtaining separated waste: metal, plastic and paper. With her father she started to collect and transport the separated waste produced by their family to these companies. Later Nastya invited her friends to join the initiative, proposing a free transportation service organized by her father. In just one year, there were already 12 families collecting separated waste along with Nastya's family.

> I want to do at least something that is in my power to improve life in this city because I live here, I communicate with these people. I think it is comfortable and pleasant to live in a nice cozy place, and not at a landfill. I try to participate as much as possible in everything.

After several months, a loosely organized group of volunteers formed around Nastya and her family. They started to organize regular street events open to the public to collect separated waste from the residents of the Kirovsk-Apatity agglomeration. Announcements of these public events are published on the local social media where volunteers are also recruited for each particular event. For Nastya, this activity became a priority in her life and she is going to continue this work: "Now I have changed my priorities, I decided to do what is vitally important for me and leave aside everything else. Weddings, birthdays, other things can wait, nature is more important for me." Nastya has a feeling of personal responsibility for the processes that are going on in her community and she tries to do everything that she can at the local level to solve the problem she is worried about. Together with other volunteers, she has also initiated negotiations with the local administration in Kirovsk, hoping to create institutionalized structural opportunities for selective collection of waste, but the negotiations are still ongoing.

Open space for youth as "third place"

One of the most common complaints of young people in northern industrial cities is the lack of modern leisure facilities. The outdated urban environment of these cities is indeed still very much rooted in the Soviet past and lacks places for informal public life. Recently, in a range of different northern cities, active young people have started to create new types of youth clubs similar to what Oldenburg has called "third place[s]" (Oldenburg 1989). Third places are informal, open to everybody and participants themselves organize the events they find interesting. Such grassroots initiatives have appeared in northern industrial cities because local youth need public gathering places which are free from the strict control and formal rules of official institutions.

In Kirovsk, a new open youth club was created by Valera (25), supported by a partner and a group of friends. Valera was born in Apatity, but moved to Kirovsk with his family when he was a child. After school, he studied at a vocational training institution and started to work as a truck driver at the mining company Apatit. Despite earning a lot of money in his young years, he soon realized that he does not like this job: "It was very difficult for me to work there. This was the Central mine and the conditions of work were the most dangerous there: this mine is high up in the mountains, there is always snow and it was always dark. And there were the most sophisticated transport vehicles in the area, it was hard to work on them." Valera quit this job, and, following a period of unemployment, found a new job as a truck driver at a subcontracting company while also starting to participate actively in community life.

> My friend inspired me to become active in the community life and I got involved in volunteer work. We were helping at different events, we watched and learned how to organize them. Then we started to propose different projects and applications and in the end, my activist's life supplanted the main job: I travelled to various youth forums and had to ask for additional unpaid vacations, so after one year I got laid off at work.

With his friend, Valera started to initiate commercial projects in the field of creative industries because he did not want to continue working for the mining industries. They bought a franchise and started to organize cinema quizzes in the region, which soon became very popular. They also continued to participate in large national youth forums and various programs for youth, acquiring the skills needed for activism, such as grant writing, project management, team building, etc. In parallel, Valera met an entrepreneur who owned a hotel in Kirovsk and who had an empty basement space at the hotel where he wanted to establish a new public space for tourists and local residents. Valera agreed to renovate the space and to create an open space there. Presently, in this small basement space, there is a combination of social and commercial activities, providing both free and paid events, such as film screenings, small concerts and festivals. Over time, a core group of young people have started to hold regular gatherings at this place, but there are always new people coming for activities of interest. For one public festival, Valera tried to get support from the Apatit company and the local administration because he needed prizes for the winners of the competition. However, he did not succeed: "There is a youth organization at the Apatit company and they are open for communication, however, they only want to support workers of the company." Support of the town-forming company is usually crucial for the existence of new social initiatives in industrial cities, but this project has continued to thrive without it, cooperating instead with other businesses and individuals interested in the development of the local community.

These cases of civic urban activism of young people demonstrate how active young people can become engaged in community development in northern industrial cities. Despite the widespread passivity in these communities, some young activists are able to organize new initiatives and projects that solve local problems and also attempt to make their industrial towns more liveable for youth.

Conclusion: staying as agency?

In the Russian Arctic, even the successful economic transformations of town-forming companies in mining towns have not been followed by correspondingly successful social transformations. The cross-generational perceptions of the localities vary considerably due to their different experiences. The older generation contributed to the rise of the towns and was actively engaged in city construction and development; through this process, they developed strong social ties and connections to the localities. In contrast, among the modern youth a feeling of decay and uncertainty pervades, even in towns with economically stable mines. The majority of young northerners dream of moving to big cities; still, many of them cannot leave due to a complex combination of social, emotional and material factors. Embedded as such in the home region and their localities, young people have to deal with the complexities of leaving/staying: first, during their transition to adulthood, and then multiple times over their life course. Often staying put in a declining industrial city is perceived as a failure, as being "stuck" in place as the world moves on (see Komu and Adams, this volume).

In the social sciences, attention is mostly paid to youth outmigration from the North and its drivers, while immobility and the agency of those who stay put are rarely investigated. This chapter contributes to the understanding of how young adults who stay in the North find their ways, experience control in their life and exercise personal agency in the particular structural conditions of northern single-industry towns. In the end, staying also requires agency, though the perception of this situation can vary significantly depending on the degree of (in)voluntariness. To better understand youth agency in the formation their futures, it is important to explore "how youth construct for themselves their actions, resistance, and imaginaries in relation to both their present situations and desired futures that are historically, socially, and culturally embedded" (DeJaeghere et al. 2016, p. 20). In this sense, this chapter is an attempt to focus on staying as an *active process*, through which young northerners make their own migration decisions and life choices responding to structural forces. This in-depth qualitative perspective on young people staying in northern industrial cities enriches the current research dominated by scholarship on outmigration from the North.

70   *Alla Bolotova*

Notes

1 Similarly, in many rural areas the outmigration of youth is an expected step in the life course of a young person (Easthope and Gabriel 2008; Abbott-Chapman et al. 2014; Nugin 2014).
2 These towns were: Revda; Kovdor; and Kirovsk.
3 This publication was supported by the WOLLIE project, funded by the Academy of Finland, decision number 314471. It is also partially based on results gained in the framework of the earlier project "Children of 1990s", funded by the Russian Science Foundation under grant no. 14-18-02136 (2014–2016).
4 I use a broad age range for defining youth, with 35 as the upper limit to follow current tendencies in Russian legislation. Recently, the State Duma of the Russian Federation adopted the first reading of the law "On youth policy in the Russian Federation" in which the upper age limit of youth is 35 (https://rg.ru/2020/11/11/gosduma-odobrila-proekt-o-povyshenii-vozrasta-molodezhi-do-35-let.html, accessed 21.12.2020).
5 E.g., see Pilkington 2012 on relationships between discursive production of place and its perception by young people in an arctic city Vorkuta.
6 For an analysis of staying as an expression of agency in the context of the northern regions see, Khlinovskaya Rockhill 2010.
7 Calculated by the author based on Rosstat 2020; Vsesoiuznaia perepis' naseleniia 1989.
8 Compared with the more temperate regions of Russia where a standard vacation period is 28 days, the minimal duration of vacation in northern regions is 52 days and employers are obliged to pay for travel expenses for their employees once every two years. The tradition to bring children for vacations to more southern regions appeared in Soviet times when northerners regularly travelled either to the Black Sea or to visit relatives in other regions.
9 Similar complaints are common in northern Finland, see Komu and Adams (in this volume).

References

Abbott-Chapman, J., R. Johnston and T. Jetson (2014) 'Rural belonging, place attachment and youth educational mobility: rural parents' views', *Rural Society, 23*(3), pp. 296–310. doi: doi:10.1080/10371656.2014.11082072.
Adams, R.-M., Allemann, L. and Tynkkynen, V.-P. (this volume). 'Youth well-being in "atomic towns": The cases of Polyarnye Zori and Pyhäjoki', in Stammler, F. and Toivanen, R. (eds) *Young people, wellbeing and placemaking in the Arctic*. London: Routledge, pp. 222–240.
Ali, S. (2007) "Go west young man: the culture of migration among Muslims in Hyderabad, India', *Journal of Ethnic and Migration Studies, 33*(1), pp. 37–58. doi: 10.1080/13691830601043489.
Béal, V., Journel, C. M., and Pala, V. S. (2017) 'From 'Black city' to 'slum city': The importance of image in Saint-Étienne', *Metropolitics*, 20 September. Available at: http://www.metropolitiques.eu/From-Black-City-to-Slum-City-The.html (Accessed: 20.06.2020).
Bernt, M. and Rink, D. (2010) "Not relevant to the system: The crisis in the backyards', *International Journal of Urban and Regional Research, 34*(3), pp. 678–685. doi: https://www.researchgate.net/deref/http%3A%2F%2Fdx.doi.org%2F10.1111%2Fj.1468-2427.2010.00985.x.

Bjarnason, T., and Thorlindsson, T. (2006) 'Should I stay or should I go? Migration expectations among youth in Icelandic fishing and farming communities', *Journal of Rural Studies*, *22*(3), pp. 290–300. doi: doi: 10.1016/j.jrurstud.2005.09.004.

Bolotova, A. and F.M. Stammler (2010) 'How the North became home: Attachment to place among industrial migrants in the Murmansk Region of Russia', in C. Southcott and L. Huskey (eds), *Migration in the Circumpolar North: Issue and contexts*. Edmonton: CCI press, pp. 193–220.

Bolotova, A. (2012) 'Loving and conquering nature: Shifting perceptions of the environment in the industrialised Russian North', *Europe-Asia Studies*, *64*(4), pp. 645–671. doi: 10.1080/09668136.2012.673248.

Bolotova, A. (2014) *Conquering nature and engaging with the environment in the Russian industrialised North*. Rovaniemi: Acta Universitatis Lapponiensis 291. University of Lapland.

Bolotova, A., Karaseva, A., and Vasilyeva, V. (2017) 'Mobility and sense of place among youth in the Russian Arctic', *Sibirica*, *16*(3), pp. 77–124. doi: 10.3167/sib.2017.160305.

Brenner, N. (2004) *New state spaces: Urban governance and the rescaling of statehood*. Oxford: Oxford University Press.

Brenner, N. (2019) *New urban spaces: Urban theory and the scale question*. Oxford: Oxford University Press.

Carling, J. (2002) 'Migration in the age of involuntary immobility: Theoretical reflections and Cape Verdean experiences', *Journal of Ethnic and Migration Studies* *28*(1), pp. 5–42. doi: 10.1080/13691830120103912.

Corbett, M. (2007) *Learning to leave: The irony of schooling in a coastal community*. Halifax: Fernwood Publishing.

Corcoran, M. (2002) 'Place attachment and community sentiment in marginalized neighbourhoods: A European case study', *Canadian Journal of Urban Research*, *11*(1), pp. 47–68. Available at: http://mural.maynoothuniversity.ie/1216/1/MCplaceattachment.pdf.

Cresswell, T. (2012) 'Mobilities II: Still', *Progress in Human Geography*, *36*(5), pp. 645–653. doi: 10.1177/0309132511423349.

DeJaeghere, J. G., Josić, J. and McCleary, K. S. (eds) (2016) *Education and youth agency: Qualitative case studies in global contexts*. New York City: Springer.

Didyk, V. (2015) 'Development challenges for a single-industry mining town in the Russian Arctic: The case of Kirovsk, Murmansk Region', *Russian Analytical Digest*, *172*, pp. 2–6. Available at: https://css.ethz.ch/content/dam/ethz/special-interest/gess/cis/center-for-securities-studies/pdfs/RAD172.pdf.

Easthope, H. and M. Gabriel (2008) 'Turbulent lives: exploring the cultural meaning of regional youth migration', *Geographical Research*, *46*(2), pp. 172–182. doi: 10.1111/j.1745-5871.2008.00508.x.

Evans, K. (2002) 'Taking control of their lives? Agency in young adult transitions in England and the new Germany', *Journal of Youth Studies.*, *5*(3), pp. 245–269. doi: 10.1080/1367626022000005965.

Evans, K. (2007) 'Concepts of bounded agency in education, work, and the personal lives of young adults', *International Journal of Psychology*, *42*(2), pp. 85–93. doi: 10.1080/00207590600991237.

Farrugia, D. (2014) 'Towards a spatialised youth sociology: the rural and the urban in times of change', *Journal of Youth Studies*, *17*(3), pp. 293–307. doi: 10.1080/13676261.2013.830700.

Fernandez-Carro, C. and Evandrou, M. (2014) 'Staying put; factors associated with ageing in one's 'lifetime home'. Insights from the European context', *Research on Ageing and Social Policy*, *2*(1), pp. 28–56. doi: 10.17583/rasp.2014.1053.

Gray, B. (2011) 'Becoming non-migrant: Lives worth waiting for', *Gender, Place and Culture*, *18*(3), pp. 417–432. doi: 10.1080/0966369X.2011.566403.

Haase, A., Rink, D., Grossmann, K., Bernt, M. and Mykhnenko, V. (2014) 'Conceptualizing urban shrinkage', *Environment and Planning A*, *46*(7), pp. 1519–1534. doi: 10.1068/a46269.

Heleniak, T. (1999) 'Outmigration and depopulation of the Russian north during the 1990s', *Post-Soviet Geography and Economics*, *40*(3), pp. 155–205. doi: doi:10.1080/10889388.1999.10641111.

Heleniak, T. and Bogoyavlensky, D. (2015) 'Arctic populations and migration', in Larsen, J. Nymand and Fondahl, G. (eds) *Arctic human development report: Regional processes and global linkages*. Copenhagen: Nordic Council of Ministers. pp. 53–104. Available at: http://norden.diva-portal.org/smash/record.jsf?pid=diva2%3A788965&dswid=5671.

Hjalm, A. (2014) 'The 'stayers': Dynamics of lifelong sedentary behaviour in an urban context', *Population, Space and Place*, *20*(6), pp. 569–580. doi: 10.1002/psp.1796.

Ho, K. C., and Douglass, M. (2008) 'Globalisation and liveable cities. Experiences in place-making in Pacific Asia', *International Development Planning Review*, *30*(3), pp. 199–213. doi: 10.3828/idpr.30.3.1.

Horváth, D. I. (2008) 'The culture of migration of rural Romanian youth', *Journal of Ethnic and Migration Studies*. Routledge, *34*(5), pp. 771–786. doi: 10.1080/13691830802106036.

Khlinovskaya Rockhill, E. (2010) 'Living in two places: Permanent transiency in the Magadan Region', *Alaska Journal of Anthropology*, *8*(2), pp. 43–61. Available at: http://www.alaskaanthropology.org/wp-content/uploads/2017/08/akanth-articles_342_v8_n2_Rockhill.pdf.

Kinder, K. (2016) *DIY Detroit: Making do in a city without services*, Minneapolis: University of Minnesota Press.

Komu, T., and Adams, R-M. (this volume) 'Not wanting to be "stuck": Exploring the role of mobility for young people's wellbeing in Northern Finland', in Stammler, F. and Toivanen, R. (eds). *Young people, wellbeing and placemaking in the Arctic*. London: Routledge, pp. 32–52.

Kong, L. (2009) 'Making sustainable creative/cultural space in Shanghai and Singapore', *Geographical Review*, *99*(1), pp. 1–22. Available at: https://ink.library.smu.edu.sg/soss_research/1701.

Kühn, M. (2015) 'Peripheralization: Theoretical concepts explaining socio-spatial inequalities', *European Planning Studies*, *23*(2), pp. 367–378. doi: 10.1080/09654313.2013.862518.

Lansbury R., Breakspear Ch. (1995) 'Closing down the mine: A tale of two communities and their responses to mining closures in Australia and Sweden', *Economic and Industrial Democracy*, *16*(2), pp. 275–289. doi: 10.1177/0143831X95162006.

Leibert, T. (2016) 'She leaves, he stays? Sex-selective migration in rural East Germany', *Journal of Rural Studies*, *43*, pp. 267–279. doi: 10.1016/j.jrurstud.2015.06.004.

Liarskaia, E., Vasil'eva, V., Karaseva, A. (2020) 'Uekhat' i ostat'sia: sotsial'naia mekhanika severnykh migratsii', in *"Deti devyanostykh" v sovremennoi rossiiskoi Arktike*. N. Vakhtin and Sh. Dudeck (eds). St. Petersburg: Izd-vo EUSPb.

Mah, A. (2012) *Industrial ruination, community, and place: Landscapes and legacies of urban decline*. Toronto: University of Toronto Press.

Mahmood, S. (2001) 'Feminist theory, embodiment and the docile agent: Some reflections on the Egyptian Islamic revival', *Cultural Anthropology, 16*(2), pp. 202–236. Available at: https://www.jstor.org/stable/656537.

Martin, S. (2009) 'The effects of female outmigration on Alaska villages', *Polar Geography, 32*(1–2), pp. 61–67. doi: 10.1080/10889370903000455.

Martinez-Fernandez C., Audirac I., Fol S. and Cunningham-Sabot, E. (2012) 'Shrinking cities: urban challenges of globalization', *International Journal of Urban and Regional Research, 36*(2), pp. 213–225. doi: 10.1111/j.1468-2427.2011.01092.x.

Mata-Codesal, D. (2015) 'Ways of staying put in Ecuador: social and embodied experiences of mobility–immobility interactions', *Journal of Ethnic and Migration Studies, 41*(14), pp. 2274–2290. doi: 10.1080/1369183X.2015.1053850.

Monogoroda Arkticheskoi zony RF: problemy i vozmozhnosti razvitiia (2016) *Analiticheskii doklad.* Moscow: IPPI. Available at: http://www.arcticandnorth.ru/Encyclopedia_Arctic/monogoroda_AZRF.pdf (Accessed: October 4 2020).

Nugin, R. (2014) 'I think they should go. Let them see something', *Journal of Rural Studies, 34*, pp. 51–64. doi: 10.1016/j.jrurstud.2014.01.003.

Oakley, K (2014) 'Good work: Rethinking cultural entrepreneurship', in Bilton, C, Cummings, S (eds) *Handbook of management and creativity.* Cheltenham: Edward Elgar Publishing, pp. 145–159.

Oldenburg, R. (1989) *The great good place: Cafes, coffee shops, community centers, beauty parlors, general stores, bars, hangouts, and how they get you through the day.* New York: Paragon House.

Pelkmans, M. (2013) 'Ruins of hope in a Kyrgyz post-industrial wasteland', *Anthropology Today, 29*(5), pp. 17–21. doi: 10.1111/1467-8322.12060.

Pilkington, H. (2012) 'Vorkuta is the capital of the world': People, place and the everyday production of the local', *Sociological Review, 60*(2), pp. 267–291. doi: doi:10.1111/j.1467-954X.2012.02073.x.

Pilyasov, A. (2013) 'Russia's policies for Arctic cities', *Russian Analytical Digest, 129*, pp. 2–4. Available at: https://css.ethz.ch/en/services/digital-library/publications/publication.html/166409.

Preece, J. (2018) 'Immobility and insecure labour markets: An active response to precarious employment', *Urban Studies, 55*(8), pp. 1783–1799. doi: 10.1177/0042098017736258.

Rasmussen, R. O. (2007) 'Polar women go south', *Journal of Nordregio, 7*(4), pp. 20–23. Available at: http://nordregio.shotcode.no/filer/Files/jon0704.pdf.

Rasmussen, R. O. (2011) *Megatrends.* Copenhagen: Nordic Council of Ministers.

Ringel, F. (2018) *Back to the postindustrial future: An ethnography of Germany's fastest-shrinking city* (Vol. 33). New York City: Berghahn Books.

Robertson, D. (2006) *Hard as the rock itself: Place and identity in the American mining town.* Boulder: University Press of Colorado.

Rosstat (2020) *Chislennost' naselenija Rossijskoj Federatsii po munitsipal'nym obrazovaniiam na 1 ianvaria 2020 goda.* Moscow: Federal'naya Sluzhba Gosudarstvennoy Statistiki (Rosstat).

Schewel, K. (2020) 'Understanding immobility: Moving beyond the mobility bias in migration studies', *International Migration Review, 54*(2), pp. 328–355. doi: 10.1177/0197918319831952.

Shashkov, V. (2004) *Spetspereselentsy v istorii Murmanskoi oblasti.* Murmansk: Maksimum.

Smith, N. (1984) *Uneven Development: Nature, capital, and the production of space.* Oxford: Blackwell.

Stockdale, A. (2002) 'Towards a typology of outmigration from peripheral areas: A Scottish case study', *International journal of population geography*, *8*(5), pp. 345–364. doi: 10.1002/ijpg.265.

Stockdale, A., and Haartsen, T. (2018) 'Editorial introduction: Putting rural stayers in the spotlight', *Population, Space and Place*, *24*(4), p. 2124. doi: 10.1002/psp.2124.

Suutarinen, T. K. (2015) 'Local natural resource curse and sustainable socio-economic development in a Russian mining community of Kovdor', *Fennia-International Journal of Geography*, *193*(1), pp. 99–116. doi: 10.11143/45316.

Vakhtin, N., Dudeck S. (eds) (2020) *"Deti devianostykh" v sovremennoi rossiiskoi Arktike*. St. Petersburg: Izd-vo EUSPb.

Vsesoiuznaia perepis' naseleniia (1989) *Chislennost' nalichnogo naseleniia soiuznykh i avtonomnykh respublik, avtonomnykh oblastei i okrugov, kraev, oblastei, raionov, gorodskikh poselenii i sel-raitsentrov*. Available at: http://www.demoscope.ru/weekly/ssp/rus89_reg2.php (Accessed: December 15 2020).

Weaver, R., Bagchi-Sen, S., Knight, J., and Frazier, A. E. (2016) *Shrinking cities: Understanding urban decline in the United States*. New York: Routledge.

Part II

Youth agency for the future

Alternatives and livelihoods

4 Towards a sustainable future of the Indigenous youth

Arctic negotiations on (im)mobility

Reetta Toivanen

Introduction: extractivist imaginaries of the Barents Sea region and the local peoples

Sustainable wellbeing in this chapter means that health, a sufficient standard of living, functioning human relationships, possibilities for self-realization, and the availability of meaningful activities are shared on a relatively equal basis among human beings now and in the future (SYKE 2020). Sustainable wellbeing is also linked to human rights, because without the universal outreach of human rights nothing is sustainable – human rights are at the core in the Sustainability Goals of the United Nations Agenda 2030 (see Toivanen and Cambou forthcoming). Sustainable wellbeing can be theorized through the concept of lived citizenship, as has been proposed by Isin (2009, 2012). According to Isin (ibid.), citizenship includes official status and membership in a society, and participation through the existing socio-political structures and practices, but also everyday being and acting in a democratic society. The idea of lived citizenship allows us to analyse the societal actions of Indigenous young people in the Arctic in a more encompassing frame than just as members of one specific society, since experienced and implemented actions exist not only in a given political system, but as everyday actions with peers and family and other social relationships (Kallio et al. 2020).

The question that I wish to tackle in this chapter is whether the decline of young, working-aged people in the Sámi homelands has a detrimental effect on Sámi languages and cultures? Furthermore, I consider a separate question: what is the connection between mobility and sustainable wellbeing? There is a strong assumption that young people living in the circumpolar region, Indigenous or not, need to choose between staying or leaving, and there is no alternative discourse on what (im)mobility might entail. This dilemma is supported by dominant contradictory narratives on the Arctic and its inhabitants.

On the one hand, sustainable development discourses often construct Indigenous peoples as part of nature, fragile and thus in need of support and protection. Consequently, Arctic peoples are imagined as people without their own agency or plans for the future (Toivanen 2019). On the other hand, businesses and state stakeholders portray the Arctic area as a periphery,

DOI: 10.4324/9781003110019-7

empty of (relevant) peoples and as a place of immense treasures to be exploited and extracted (ibid.). Steinberg et al. (2015) argue that the political tendency to imagine the circumpolar North as empty has been especially pronounced when there has been a strong interest in the local natural resources. In this narrative, the people living in the Arctic region play no role in the future of the area because they are rendered irrelevant. This is visible in the political strategies on the development of the Arctic that often ignore the fact that there are real people living in that area. As an example, Finland is one of the countries that has devised a national strategy to implement UN Agenda 2030, yet it is revealing that the *Government Report on the 2030 Agenda for Sustainable Development—Sustainable Development in Finland* (Prime Minister's Office 2017) makes no mention of the Sámi Indigenous peoples living in the country. Neither does the document *Opportunities for Finland* (a joint outlook of the Permanent Secretaries of the ministries on the key questions for the upcoming 2019–2023 government term) (Finnish Government 2019; see also Toivanen and Cambou forthcoming). At the same time, there are again a growing number of plans on how to exploit the Arctic environment for the purposes of national and even global wellbeing (e.g. the building of windmills in Sámi areas, building a railway line through the Sámi homelands; ibid.).

The local populations and Indigenous peoples are not part of the majoritarian narratives on the Arctic region (Ryall et al. 2010; Toivanen 2019). There are signs of politics changing in this regard and the Sámi cannot be totally ignored in development plans. Still, Indigenous ecological knowledge is too seldom included in decision-making in the Arctic (Mustonen and Feodoroff 2018; see also Casi et al. forthcoming).

Many of those who move south to go to secondary school, universities and applied educational institutes stay away permanently and build new lives in the urban centres. They do not, however, as assumed in much of the literature on minorities and language and culture loss, necessarily abandon Sámi culture, languages and lands (Paksuniemi and Keskitalo 2019). Indeed, contrary to the fears of the older generation and expectation of the general narrative, they also remain closely connected to their Sámi lands, their families, cultures and languages (Toivanen and Fabritius 2020). The young Sámi find creative ways to challenge the dominant societal discourses that have for decades predicted the end of Sámi cultures and languages, as will be shown later in this chapter.

This chapter is based on observation and interview materials from the years 2011–2020[1] in three Arctic municipalities in Norway, Finland and Russia.[2] The interviews were conducted with young people, families and different local stakeholders, such as teachers, local politicians and municipal workers. The interviews and extracts of field notes have been analysed with the Atlas.ti qualitative data analysis and research software program in order to identify the common themes and topics of interest through a discourse analytical lens (Wodak 2015). Using critical discourse analysis, the aim has been to understand how the interviewees construct their being in the world and their views on how societies work (Fairclough 2003, p. 203).

The research material has been used for several other scientific articles that have focused on wellbeing, solidarity and sustainability. For this chapter, a new approach was taken on the material and what it tells about youth in the Arctic area. I have chosen here to concentrate on the research carried out in Finland, but there are similarities in ways of being young across the Sámi home areas in all of the Sámi homeland, Sápmi. The identified discourses are analysed within the framework of wellbeing, defined as sustainable wellbeing (Kjell 2011), which is enacted through lived citizenship (Isin 2012). Moreover, (im)mobility theories are applied in order to consider what the dichotomy "leaving versus staying" actually tells about youth mobility in the Arctic region.

The European Arctic and its peoples

The reason for the loss of inhabitants in the area is that the young and people in employment age move away and population concentrates in cities.

(Tennberg et al. 2017, 143)

The Barents region is the most populated area in the Arctic. Also in terms of infrastructure, including roads, electricity and internet networks, the area is well developed. However, during the last two decades, population growth in the Arctic has occurred in Alaska, Iceland, the Canadian Arctic and West Siberia, whilst the Barents region has mostly been experiencing population decline (Emelyanova and Rautio 2016). By January 2015, approximately 5.1 million people permanently resided in the area, a fifth lower than was recorded about two decades ago (6.5 million in 1990; Emelyanova and Rautio 2016, p. 5; Tennberg et al. 2017). The losses were particularly noticeable in the north-west corner of the Russian Federation, whereas in Finnish Lapland, for example, the population had declined only moderately. The northern Norwegian population remained roughly at the same level (Emelyanova and Rautio 2016). In the 1990s, most of the northern regions of Norway, Sweden and Finland—and also throughout the 2000s the Russian North—had slower population growth or faster decline than the rest of their respective countries (Emelyanova and Rautio 2016, p. 5).

One of the peoples traditionally living in the Barents region in the Arctic are the Sámi Indigenous people. The number of Sámi is smaller than that of the majority populations in most of the northern municipalities, and thus they share living space with other local peoples (different minorities and other people who have lived in the territories for several generations)[3] and newcomers. According to some estimates, there are approximately 75,000 to 90,000 Sámi people: ca. 40,000 in Norway, 25,000 in Sweden, 11,000 in Finland, and 2,000 in Russia on the Kola peninsula. The Sámi share closely related languages and many cultural features (Kulonen et al. 2005; Pietikäinen 2010). Politically, they have worked together since 1956 in the Sámiráđđi (Sámi Council, earlier Nordic Sámi Council) to strengthen their voices vis-à-vis the majority states of Norway, Sweden, Finland (see Toivanen 2003) and, in recent decades, also Russia after a long period of denying the Sámi cultural rights and languages (see Kotljarchuk 2019). Certain occupations have been defined as traditional Sámi livelihoods, such as reindeer herding, fishing, and hunting and gathering (Tennberg et al. 2017, p. 50). Alongside these more traditional professions, there are many new occupations in the fields of tourism and education. Some jobs actually pose a danger for Sámi futures because they negatively impact the environment, such as working in the mining or forest industries.

The fact that the number of people living in the northernmost municipalities of the four countries is declining is a worrisome phenomenon (see also Simakova et al. and Ivanova et al., both in this volume). Because no ethnic statistics are available, the effects of this on Sámi people requires more research (AMAP 2017; see also Tennberg et al. 2017). To explain this trend in terms of Finland, the reasons for young people to leave are rather obvious. First, there is a lack of educational opportunities. There are only three upper secondary schools in the home area of Sámi peoples where it is possible to continue education after comprehensive education. There are five vocational and two higher (one university) educational institutions in Finnish Lapland, but none in the Sámi homeland area. The Sámi Education Institute (SAKK) is the only vocational college in the Sámi home area and its educational opportunities are limited to a few professions. Second, there is a lack of employment opportunities. Thus, finding work is one reason to leave. The unemployment rate can be over 19% in some municipalities in Lapland, compared to the Finnish average of 7.6% (Statistics Finland 2020)[4]. There is a strong narrative that if there were better opportunities for education and employment, young people would not need to leave (see e.g., Tennberg et al. 2017) and they would stay instead. But is there necessarily an either/or here, with only two alternatives to choose from? I will argue that being (im)mobile and seeking sustainable wellbeing go hand in hand, leaving space for a multitude of other alternatives then the two opposites.

Based on the current statistics, it is probable that the number of Sámi Indigenous peoples living outside Sápmi—namely, their home territories in all of the four countries in which they reside—continues to increase (Nyseth and Pedersen 2014; Heikkilä et al. 2019). This means that in the whole area of Sápmi, the population is aging rapidly. According to Tennberg et al. (2017, p. 50), this development could be reversed if the young and educated Sámi population were to find employment that matches their education. To achieve this, they emphasize, there is a need to ensure that the traditional livelihoods remain a viable option. The fact that Sámi peoples are moving away from their home territory is not a just a simple act of mobility. Instead, it leads to a negative impact on the general availability of Sámi language instruction in the schools and services in Sámi languages; the less language users, the less language services (Laakso et al. 2016). Furthermore, it also weakens the general right to practice these livelihoods. The rights to practice a traditional livelihood, such as reindeer herding, fishing and hunting with traditional traps, are tied to the place of residence (Tennberg et al. 2017, p. 50). This means that when the young people move away from the homelands, they lose their rights to many traditional livelihoods.

Which professions do the young people fill? In Finland in 2020, there were still 4,300 registered reindeer herders, among whom around 900 are under 25.[5] The high number can partly be explained by the fact that in Sámi families, also children receive their first own reindeer and may appear in the statistics, even though they may not practice herding as a profession.[6] When looking at the statistics in Finland more closely, however, we can note that despite a decrease in the number of reindeer herders, it is still the profession that has the biggest number of Sámi speakers: 13% of Sámi speakers worked in a domestic situation involving animal husbandry—almost all with reindeer.[7] In addition to reindeer herding, there only a few people involved in other livelihoods that are considered traditional Sámi professions. Only a few Sámi fishers are working full-time and none of them are young, according to Statistics Finland. Nor are there professional full-time Sámi hunters or handicraft makers found in Statistics Finland in year 2016.[8]

Agriculture, forestry and fishing today employ 50% less people than in 1987. In particular, the number of people employed in reindeer herding has diminished significantly. At the same time, the number of people working in social and health services has increased in the last 30 years by over 80%. Also, the number of people working in the field of education and who use Sámi as their mother tongue has increased by 60% and in the field of public administration by 45%.[9] One explanation is the Sámi language law (1086/2003), which both encouraged and enabled the learning of Sámi languages (i.e. in-service training) and provided opportunities to use Sámi languages.

If we still look at the Finnish area of Sápmi, people with a university education make up 23.8% of the population in Inari, 23.9% in Sodankylä, 21.3% in Enontekiö and 26.8% in Utsjoki (Statistics Finland, 2020). This must be compared with the Finnish average of 41%, and the Organisation for Economic Co-operation and Development (OECD) average of 44%.[10] The difference is significant and tells about the problems of highly educated people to find work in Finnish Lapland. Those who are highly educated take positions in schools, kindergartens, administration and media.

Sustainable wellbeing and (im)mobilities

The youth in the study from the rural regions in the Nordic Arctic will during the next 10–15 years be engaged in getting an education or establishing as part of the labour market. These young people expect to be mobile and move from their place of origin in order to achieve future dreams. Only a small portion of the participating young people expect to be living in their place of upbringing during the next ten years.

(Karlsdóttir and Jungsberg 2015, 12)

Above is a citation from a publication that was produced as part of the project Foresight Analysis for Sustainable Regional Development in the Nordic Arctic, commissioned by the Nordic Working Group for Sustainable Regional Development in the Nordic Arctic (see also Karlsdóttir et al. 2017). In a way, it represents a rather hegemonic narrative of Arctic youth who move away from the home territories, voluntarily or not. The emphasis is on youth leaving, and the question of what to do about it dominates the discourse in the existing policy papers and reports.

Quite to the contrary of the citation above, my research material does not support a discourse that presupposes that achieving "future dreams" means necessarily moving and staying away from homelands. The dominant discourse furthermore suggests that remaining in a certain space and place—Indigenous homelands—would be the prerequisite for the healthy and sustainable wellbeing of Indigenous youth. To the contrary, young people in my research seemed to view mobility and wellbeing as going well hand in hand. Being mobile, or changing place, does not in their view mean the abandoning of their Sámi home. For them, the Sámi home is not only a specific area of land but more of a relationship (Keskitalo 2019; see also Virtanen 2012 on Amazonian Indigenous youth).

The term (im)mobility refers to something potentially in between being mobile and not being mobile (Zickgraf 2019). It takes into account the fact that some people may wish to be mobile and move, whereas others with the same sentiments will need to stay because of the multifarious constraints they face. Such constraints include a lack of resources, legal hindrances to obtain visas to other countries, or even personal reasons for not being able to go (Suliman et al. 2019). It also takes into account that some may not be willing to move but are left without any option, for example, due to natural catastrophes that force their movement to a new place (Carling and Schewel 2018). Finally, there are obviously also people who want to stay—and do in fact stay. Thus, (im)mobility is used in this chapter as a term to describe the multifarious aspects of moving or not moving and all the negotiations that may involve.

When it comes to Indigenous youth in the Arctic, there is a strong narrative suggesting that the youth either want to leave or have no option to stay (see Komu and Adams in this volume on their research on Northern culture surrounding migration). As noted above, this is supported by statistical information showing how, over recent decades, the northern municipalities have experienced declines in population. Whereas the demographic information needs to be taken as a given, the question is still whether the decrease of young working-aged people from the Sámi homelands has a detrimental effect on Sámi languages and cultures. What is the connection between mobility and sustainable wellbeing? By definition, sustainable wellbeing is a comprehensive and all-encompassing concept, which requires that the person's whole identity is respected and allowed to be and that it can develop in circumstances that are safe and supporting. Moving or living outside of one's safe "home" would seem to endanger this integrity.

I try to answer these questions with quotations from my interview material from Finland. Therefore, it is worthwhile mentioning that in Finland there are three language groups of Sámi: Inari, Skolt and Northern Sámi. They have different languages and cultural traditions, but also quite differing experiences of forced assimilation. The situations in Norway and Russia vary in many ways, for example, in respect to educational alternatives and other possible life paths in the Sámi traditional homelands. In Norway, the state attracts university students back to the North with financial support (see Marjomaa 2012). In Russia, there is a lack of workforce and no problem of unemployment (see Stammler and Khlinovskaya 2011 on the Russian Arctic). At the same time, these areas share many similarities, such as the fact that the majority of Sámi live outside the traditional territory and no longer speak a Sámi language.

The following extract from an interview shows how a Sámi who grew up far away from the homeland connects with their Sámi identity.

INTERVIEWER (I): Have you ever thought that you have at some point in your life been, like, that you have not so much thought about being Sámi? Or am I wrong? Or have you always known very strongly that you are a Sámi?

RESPONDENT (R): No. I did not understand it.

I: When did you, so to say, wake up to it or...?

R: Hmm, it was my cousin who at some point asked me to join the City Sámi Association's activities. It was then. But, of course, I had always had that contact to the culture. I have grown up as a child of a Sámi in the capital city region and our culture has travelled with us there. Many of our relatives have moved there, cousin's families and stuff. We kept a lot of contact and Sáminess was also there when we went for holidays to the North. In the North, in Utsjoki and Inari, there we have relatives and so forth. But it was a certain kind of identity awakening when I started being myself active in the City Sámi.

(Interview 041)

This person has grown up in the city. So have many of her relatives, and they have kept in contact with each other in the city. They have also travelled to the North to visit other relatives during holidays. She feels that the Sámi culture has been there all the time, despite the distance to the homeland. At the same time, only after she enacted her Sámi identity through participation in a Sámi association did it really become her conscious identity. As Isin (2009) points out, enacting and participation—the experience of being—bring true belonging. The distance and being mobile did not take away the experience of belonging. But it was, in particular, participation in the City Sámi activities that brought conscious identification as Sámi.

The following extract from another interview shows that young people have no need for an anachronistic understanding of Indigenous culture as something that does not change.

INTERVIEWER (I): But there is a real outcry that over 70% of Sámi children are born outside of Sámi home territory.

RESPONDENT (R): But is it true?

I: It is just [...]

R: Official?

I: Yes, there is study by the Ombud's Office on Minorities, and the Sámi Parliament has a study on it, too.

R: They most probably come. They are estimates [...]

I: [...] over 60% of Sámi Parliament's voters live outside [the Sámi home territory] [...] and then one thinks that a child who was born in the eastern part of Helsinki and whose mother might have spoken Sámi to her and maybe she received two hours a week of mother-tongue instruction in the school, could that person have a rather eastern Helsinki kind of relationship to reindeer?

R: It may be that way. Exactly that also being Sámi is also changing. It must. Because most probably a person who grew up in Helsinki thinks in a different way than a person who grew up in Angeli village. There is a difference. But there must be a reason that people who grow up in cities do what they have until now done, those who identify with Sámi identity, that they come here [to the home territory] for a while and some even stay forever. Somehow they want to come and the relationship to family is so strong that even though they would live in eastern Helsinki, their grandparents and other relatives are still in their lives.

(Interview 045)

For this person, the question is not where one is born or where one grows up, because it is the relationship with the family that makes that person. Identity is always relational, as Anna Tsing (2007) points out. As the young people tell, it is in these webs of relationships where the Sámi identity is maintained, regardless of place (Casi et al. forthcoming). Living in a city and maintaining a relationship with relatives creates another way of being Sámi. It does not undermine the identity or make it in any way weaker, because that relationship ties the person to the Sámi space. Many of the young people whom I talked to spoke about moving away and coming back and then perhaps moving again (on (im)mobilities that continue, see Zickgraf 2019). For instance, it was for school that one had to move in the first place, since the schools are in village centres or cities. It was quite natural to move and, for many young people, often even a freeing experience to live outside the sight of their parents and relatives. At the same time, the places—the Sámi lands—do not go anywhere. There is no need to fear that they could not return one day if they wanted to. The sustainable wellbeing of the interviewed youth reflects that moving does not entail an interruption of being, but that being mobile always includes the real option of immobility and staying (Zickgraf 2019).

When considering the future of Indigenous cultures and the viability of languages and cultural features, the youth seemed even more optimistic than the older generation. Some of the older people were worried that the Sámi who left and went to study and work in the cities would become incapable of continuing the Sámi traditions. Some young people shared these worries as well. Their fears were not only connected to the question of youth mobility, but much more about the larger societal questions of climate change and extractivism in Sámi areas. The following extract serves to illustrate the nature of these fears.

INTERVIEWER: How is this from the Sámi culture point of view? How do you see the connection to reindeer herding? [...] Does the extracting of natural resource threaten the Sámi culture?

RESPONDENT: Yes, it threatens because the reindeer is the cornerstone of Sámi culture. It carries with it from generation to generation knowledge, language that as such contains its own valuable knowledge. And the reindeer has so many dimensions, such as food and handicraft, and then it is such as nature. That I will definitely say, that reindeer herders have a very specific relationship to nature, so different to, for example, some Sámi teacher. And I find it worrisome that our youth can no longer move in the forest. They cannot read the signs of nature. They cannot catch fish. They don't know where reindeer go. That reindeer herding is our last link to nature and how to utilize it in a sustainable manner.

(Interview 008)

Another young person was much more optimistic about the youth's capabilities to maintain and develop Sámi cultures. This following extract illustrates a certain kind of new wave, a pluralistic attitude among some young people, especially those who are involved in the arts.

INTERVIEWER: What do you think, how do different age groups think about the future of Sámi culture?

RESPONDENT: Hmm, it has really been a strong revival in the youth culture. For example, there are new artists that have come up and even on the radio there has been a show for young audiences. It is really great that in this Sámi youth culture the ethos is that all three Sámi languages spoken in Finland and all the three groups are given visibility, and for example on this radio programme all Sámi languages are spoken. It is like, I hope that a certain kind of multicultural Sáminess would become the norm here, that we can and we must even speak all Sámi languages.

(Interview 024)

This person sees Sámi culture as a rich vessel from which new ideas, music and arts grow and that all Sámi groups together can make a real difference in strengthening the culture for the future, regardless of place. For example, the radio is accessible anywhere and it brings all the different Sámi language groups together (on revitalization and radio and social media, see Edygarova 2016).

Another young person underlined that, whereas there is so much talk about the Sámi cultures disappearing and declining when young people move to cities, he can only see that even very endangered peoples, such as Skolt Sámi, are experiencing a renaissance. This reflects a common feeling throughout the Arctic that young people are optimistic about the future of Sámi cultures and languages, and they see that the new types of social media have come to support the languages' vitality in meaningful ways, such that even people who had earlier abandoned their Sámi identity were eagerly coming back.

INTERVIEWER: How do young people relate—we can now think that you represent that generation—how do young people relate to cultural continuity?

RESPONDENT: In my view, youth think quite positively about it. It has now so much been coming, you know... Skolt Sámi culture has reached such popularity among young people that even people who never ever admitted that they are Skolt Sámi, now they so much want to be Skolt. And the traditional costumes of Skolt Sámi, even those are popular again. Soon everyone wears them, even though there was the time that nobody put them on. And so it is with the Skolt language question, that terribly many language courses have been organized and people are so interested. I don't know why they are, but it is for sure a positive thing that they are.

(Interview 010)

When asked whether the positive attitude of the young is catching on with the older generations, one Sámi woman answered:

Old people are joining and it's kind of the most wonderful of all when you witness that, old people, the elderly, who have seen all this that has happened for decades... Then when you see the children, grandchildren, grand-grandchildren suddenly start to speak Sámi [...] you can't be anything else than pleased.

(Interview 015).[11]

This ongoing urbanization is something that Indigenous locals argue is a cause for concern, and the reason for this worrying development involves the economic and infrastructural changes of modern society. One interviewee stated: "Well, of course, they leave. They have to leave when aiming for higher education, and when they get new jobs there in that place, well, then they stay there" (Interview 039). In this way, leaving is portrayed as forced by the circumstances.

Another interviewee expressed similar concerns when arguing that youth who are active in cultural maintenance are needed, but that even though "many of them would like to stay here in their region of birth" [...] "the young will leave, they leave for daily bread, they have to!" (Interview 006). In this way, Sámi culture is constructed as intrinsically bound up with the Sámi homelands, with youth leaving as forced and a threat to Sámi culture being sustained.

Whereas young people do not consider leaving as an act that separates them from Sámi culture, the concept of land is very strong for many Indigenous youth. This relationship does not seem to become weakened, even for those who have chosen or felt forced to move from their place of birth in the North. One young Sámi woman explained in an explicit way what was witnessed during the fieldwork:

> But like, for me it's really important, I have moved an insane amount of times. And still I know that when I go to [my home village] and I put my hand on the land, then I know that this is it, there my home is. It's really important for me to have that place, where I know that there they have watched the landscapes a hundred years ago, there they watched them two hundred years ago, there they watched the landscapes a thousand years ago. It is really important. Like, I define Sáminess through, like the same what it means to be Indigenous. It is the same as being Sámi. It is what is important to me, what I hold valuable, what my values are, what my worldview is. Through that my Sáminess is built. It's not built from anything else. And from that I know where the roots are from which I have been raised.
>
> (Interview 040)

This person was born and raised in her early years in Sápmi, and the connection remains very strong despite the fact that she has moved so many times—and even though she may not end up living her life in the village and on the land that she connects with so strongly. Feeling a strong connection to the homeland, she does not experience leaving as a definitive abandonment of Sámi culture.

Many envision themselves as still living in their place of birth but simultaneously commuting to work in some urban area, or living in an urban area but frequently visiting their homeland and maintaining ties through social media, as well as a strong place identity connected to the traditional land. Social media, distant learning, easy travelling and different cultural festivals in the various countries are supporting young people's connection to Sámi languages and the formation of shared Indigenous identities. In scholarly attempts to decolonize notions of place, similar observations have been made among Indigenous urban youth in other places as well (see Virtanen 2012 on Amazonian youth and Greenop 2009 on Indigenous youth in Queensland, Australia). Relationships replace the concrete concept of land as the basis for identity.

Conclusions: (im)mobility allowing for sustainable futures

I began this chapter by stating that there is a strong narrative that if the young people living in the Indigenous Sápmi territories were to have better opportunities for education and employment, they would not need to leave. The other strong narrative is that when they leave, they endanger the sustainability of Sámi languages and culture (and livelihoods). I asked whether the mobility of young people leads to erosion of Sámi languages and culture and if there is necessarily an either/or of two alternatives to choose from? I analysed my research materials and arrived at another conclusion, that being (im)mobile and seeking sustainable wellbeing can go hand in hand. There is enough space in young peoples' lives for a multitude of other alternatives then the two opposites. Thus, binary thinking—which, according to Siraj and Bal (2017, 398), "counterpoises mobility and immobility as two antagonistic concepts"—does not seem to apply to young Sámi. Instead, I argue that mobility and immobility are intrinsically related and their relationship is asymmetrical.

Recent research has shown that young people find new, creative ways of voicing and experiencing Sámi identity (Mathisen et al. 2017; Toivanen and Fabritius 2020; Joona and Keskitalo, this volume). They are fluent in expanding its discursive and geographical reach, thereby transcending the restrictive dichotomies of nature and culture, tradition and modernity, and centre and periphery.

I believe that it is important, as Wiegel et al. (2019) argue, to develop a grounded understanding of the diverse ways in which people cope with migration pressures (as well as educational pressures, desires for employment, or environmental change). This means that in addition to people's capacities for (im)mobilities, we also need to study their desires to engage in particular (im)mobility practices. What I mean here is that (im)mobility practices, real or imagined, involve young people, especially young women, moving from Lapland to southern cities, so that the Sámi culture and languages get lost. According to Nyseth and Pedersen (2014), it is precisely the moving to urban areas that has been a symbol of assimilation, cultural loss and losing one's roots. My study indicates that the young people of today's generation do not want to engage in (im)mobility practices that would cut their roots and souls from their Indigenous identities. They locate Sáminess in the relational webs around them, independent of the place where they reside.

Notes

1 This chapter was mainly funded by the Academy of Finland; see the acknowledgements section.
2 The interviews (ca. 260) are all transliterated, coded and anonymized. The interviewed were mostly carried out in the majority languages of the respective countries.
3 The question of who is Sámi is full of political tensions in Finland but not in other countries. See Heinämäki et al. 2017.
4 This contrasts with the Russian Arctic, where there is a shortage of labor, and yet people still leave, see Ivanova et al. and Bolotova, both in this volume.
5 https://paliskunnat.fi/poro/poronhoito/poromiehen-ammatti/poronuoret/
6 https://paliskunnat.fi/py/materiaalit/tilastot/poronomistajat and https://paliskunnat.fi/poro/poronhoito/poromiehen-ammatti/poronuoret/.

7 In Finland, the reindeer counts as a domestic animal—even though in the Sámi home area it mostly lives in freedom.
8 c, https://www.tilastokeskus.fi/tietotrendit/blogit/2019/hyvaa-saamelaisten-kansallispaivaa/).
9 Ibid, fn. 7.
10 Valtioneuvosto 2019, https://valtioneuvosto.fi/-//1410845/oecd-vertailu-suomessa-kilpailu-korkeakoulupaikoista-on-kovaa.
11 There are also many older people who rejected their own Sámi identity due to the strong assimilation policies and practices and who cannot understand why their children or grandchildren would voluntarily subscribe to a minority identity (see, e.g., Toivanen and Fabritius 2020; Sarivaara 2016; Grenoble and Whaley 1998).

References

AMAP (2017) *Adaptation actions for a changing Arctic: Perspectives from the Barents Area*. Oslo: Arctic Monitoring and Assessment Programme (AMAP). Available at: https://www.amap.no/documents/download/2981/inline.

Bolotova, A. (this volume) 'Leaving or Staying? Youth Agency and the Livability of Industrial Towns in the Russian Arctic', in Stammler, F. and Toivanen, R. (eds) *Young people, wellbeing and placemaking in the Arctic*. London: Routledge, pp. 53–76.

Carling, J. and Schewel, K. (2018) 'Revisiting aspiration and ability in international migration', *Journal of Ethnic and Migration Studies*, 44(6), pp. 945–963. doi: 10.1080/1369183X.2017.1384146.

Casi, C., Guttorm, H. and Virtanen, P. K. (forthcoming) 'Traditional ecological knowledge', in Krieg, C. P. and Toivanen, R. (eds.) *Situating sustainability: A handbook of contexts and concepts*. Helsinki: Helsinki University Press, pp. 202–216.

Edygarova, S. (2016) 'Standard language ideology and minority languages: The case of the permian languages', in Toivanen, R. and Saarikivi, J. (eds.) *Linguistic genocide or superdiversity? New and old language diversities. Series on linguistic diversity and language rights*. Blue Ridge Summit, PA: Multilingual Matters, pp. 326–352. doi: 10.21832/9781783096060.

Emelyanova, A. and Rautio, A. (2016) 'Population diversification in demographics, health, and living environments: The Barents Region in review', *Nordia Geographical Publications*, 45(2), pp. 3–18. Available at: https://nordia.journal.fi/article/view/64852.

Fairclough, N. (2003) *Analyzing discourse: Textual analysis for social research*. London: Routledge.

Finnish Government (2019) Opportunities for Finland. *Publications of the Finnish Government 2019*, 3 Available at: http://urn.fi/URN:ISBN:978-952-287-694-2.

Greenop, K. (2009) '*Place meaning, attachment and identity in contemporary Indigenous Inala, Queensland*'. Conference paper at Perspectives on urban life: Connections and reconnections, 28 September–2 October 2009, Australian National University, Australian Institute of Aboriginal and Torres Strait Islander Studies, Canberra.

Grenoble, L. and Whaley, L. (1998) *Endangered languages: Language loss and community response*. Cambridge: Cambridge University Press.

Heikkilä, L., Laiti-Hedemäki, E. and Miettunen, T. (2019) *Buorre eallin gávpogis: saamelaisten hyvä elämä ja hyvinvointipalvelut kaupungissa* [Good life of Sámi and wellbeing services in the city]. Rovaniemi: Lapin Yliopisto.

Heinämäki, L., Allard, C., Kirchner, S., Xanthaki, A., Valkonen, S., Mörkenstam, U., Bankes, N., Ruru, J., Gilbert, J., Selle, P., Simpson, A. and Olsén, L. (2017)

Saamelaisten oikeuksien toteutuminen: kansainvälinen oikeusvertaileva tutkimus. [Actualizing Sámi Rights: International Comparative Research] Helsinki: Valtioneuvoston kanslia (Valtioneuvoston selvitys- ja tutkimustoiminnan julkaisu-sarja, nro 4/2017).

Isin, E. F. (2009) 'Citizenship in the flux: The figure of the activist citizen', *Subjectivity*, *29*(1), pp. 367–388. doi: 10.1057/sub.2009.25.

Isin, E. F. (2012) *Citizens without frontiers.* New York and London: Bloomsbury Publishing Plc.

Ivanova, A., Oglezneva, T. and Stammler, F. (this volume) 'Youth law, policies and their implementation in the Russian Arctic', in Stammler, F. and Toivanen, R. (eds) *Young people, wellbeing and placemaking in the Arctic.* London: Routledge, pp. 147–169.

Joona, T. and Keskitalo, P. (this volume) 'Youths' and their guardians' prospects of reindeer husbandry in Finland', in Stammler, F. and Toivanen, R. (eds) *Young people, wellbeing and placemaking in the Arctic.* London: Routledge, pp. 93–119.

Kallio, K. P., Wood, B. E. and Häkli, J. (2020) 'Lived citizenship: conceptualising an emerging field', *Citizenship Studies*, *24*(6), pp. 713–729. doi: 10.1080/ 13621025.2020.1739227.

Karlsdóttir, A. and Jungsberg, L. (2015) *Nordic Arctic youth future perspectives.* Stockholm: Nordregio.

Karlsdóttir, A., Olsen, L. S., Harbo, L. G., Jungsberg, L. and Rasmussen, R. O. (2017) 'Future regional development policy for the Nordic Arctic: Foresight analysis 2013–2016', *Nordregio Report 2017*, p. 1. Available at: http://norden.diva-portal. org/smash/get/diva2:1069494/FULLTEXT01.pdf (Accessed January 2021).

Keskitalo, P. (2019) 'Nomadic narratives of Sámi people's migration in historic and modern times', in Uusiautti, S. and Yeasmin, N. (eds.) *Human migration in the Arctic.* Singapore: Palgrave Macmillan, pp. 31–65. doi: 10.1007/978-981-13-6561-4_3.

Kjell, O. N. E. (2011) 'Sustainable well-being: A potential synergy between sustain-ability and well-being research', *Review of General Psychology*, *15*(3), pp. 255–266. doi: 10.1037/a0024603.

Komu, T. and Adams, R.-M. (this volume) 'Not wanting to be "stuck": Exploring the role of mobility for young people's wellbeing in northern Finland', in Stammler, F. and Toivanen, R. (eds) *Young people, wellbeing and placemaking in the Arctic.* London: Routledge, pp. 32–52.

Kotljarchuk, A. (2019) 'Indigenous people, vulnerability and the security dilemma: Sami school education on the Kola Peninsula, 1917–1991', in Kortekangas, O., Keskitalo, P., Nyyssönen, J., Kotljarchuk, A., Paksuniemi, M. and Sjögren, D. (eds.) *Sámi educational history in a comparative international perspective.* Cham: Palgrave Macmillan, pp. 63–82.

Kulonen, U.-M., Seurujärvi-Kari, I. and Pulkkinen, R. H. (2005) *The Saami: A cul-tural encyclopaedia.* Helsinki: Suomalaisen Kirjallisuuden Seura (Suomalaisen Kirjallisuuden Seuran toimituksia, no. 925).

Laakso, J., Sarhimaa, A., Spiliopoulou Åkermark, S. and Toivanen, R. (2016) *Towards openly multilingual policies and practices. Assessing minority language maintenance across Europe.* Bristol: Multilingual Matters.

Marjomaa, M. (2012) 'North Sámi in Norway: An overview of a language in context', *Working papers in European Language Diversity*, *17*. Available at: https://www.oulu. fi/sites/default/files/content/Giellagas_Marjomaa_NorthSamiInNorway.pdf.

Mathisen, L., Carlsson, E. and Sletterød, N. A. (2017) 'Sámi identity and visions of preferred futures: Experiences among youth in Finnmark and Trøndelag, Norway', *The Northern Review*, *45*, pp. 113–139. doi: 10.22584/nr45.2017.007.

Mustonen, T. and Feodoroff, P. (with the Skolt Sámi Fishermen of Sevettijärvi) (2018) 'Skolt Sámi and Atlantic Salmon Collaborative Management of Näätämö Watershed, Finland as a case of Indigenous evaluation and knowledge in the Eurasian Arctic', *Indigenous Evaluation – New Directions for Evaluation, 159*, pp. 107–119. doi: 10.1002/ev.20334.

Nyseth, T. and Pedersen, P. (2014) 'Urban Sámi identities in Scandinavia: Hybridities, ambivalences and cultural innovation', *Acta Borealia, 31*(2), pp. 131–151. doi: 10.1080/08003831.2014.967976.

Paksuniemi, M. and Keskitalo, P. (2019) 'Christian morality and enlightenment to the natural child: Third-sector education in a children's home in Northern Finland (1907–1947)', in Kortekangas, O., Keskitalo, P., Nyyssönen, J., Kotljarchuk, A., Paksuniemi, M. and Sjögren, D. (eds.) *Sámi Educational History in a Comparative International Perspective*. Cham: Palgrave Macmillan, pp. 161–185. doi: 10.1007/978-3-030-24112-4_10.

Pietikäinen, S. (2010) 'Sámi language mobility: scales and discourses of multilingualism in a polycentric environment', *International Journal of the Sociology of Language, 2010*(202), pp. 79–101. doi: 10.1515/ijsl.2010.015.

Prime Minister's Office (2017) *Prime Minister's office government report on the 2030 agenda for sustainable development – Sustainable development in Finland – Long-term, coherent and inclusive action*. Helsinki: Prime Minister's Office Publications (11/2017).

Ryall, A., Schimanski, J. and Wærp, H. H. (2010) 'Arctic Discourses: An introduction', in Ryall, A., Schimanski, J. and Wærp, H. H. (eds) *Arctic Discourses*. Cambridge: Cambridge Scholars Publishing, pp. ix–xxiii.

Sarivaara, E. (2016) Emergent Sámi identities – From assimilation towards revitalization, in Toivanen, R. and Saarikivi, J. (eds.) *Linguistic genocide or superdiversity? New and old language diversities*. Bristol: Multilingual Matters, pp. 357–404.

Simakova, A., Pitukhina, M. and Ivanova, A. (this volume) 'Motives for migrating among youth in Russian Arctic industrial towns', in Stammler, F. and Toivanen, R. (eds) *Young people, wellbeing and placemaking in the Arctic*. London: Routledge, pp. 17–31.

Siraj, N. and Bal, E. (2017) 'Hunger has brought us into this jungle: Understanding mobility and immobility of Bengali immigrants in the Chittagong Hills of Bangladesh', *Social Identities, 23*(4), pp. 396–412. doi: 10.1080/13504630.2017.1281443.

Stammler, F. and Khlinovskaya, E. (2011) 'Einmal Erde und zurück: Bevölkerungsbewegungen in Russland's hohem Norden', *Osteuropa, 61*(2–3/2011), pp. 347–369. Availabe at: http://www.zeitschrift-osteuropa.de/hefte/2011/2-3/einmal-erde-und-zurueck/.

Statistics Finland (2019) Municipal Key figures 2019. Available at: https://pxnet2.stat.fi/PXWeb/pxweb/en/Kuntien_avainluvut/.

Statistics Finland (2020) *Työvoimatutkimus 2020*. Available at: http://www.stat.fi/til/tyti/2020/12.

Steinberg, P. E., Tasch, J. and Gerhardt, H. (2015) *Contesting the Arctic: Politics and imaginaries in the circumpolar North*. London: I.B. Tauris.

Suliman, S., Farbotko, C., Ransan-Cooper, H., McNamara, K. E., Thornton, F., McMichael, C. and Kitara, T. (2019) 'Indigenous (im)mobilities in the Anthropocene', *Mobilities, 14*(3), pp. 1–21. doi: 10.1080/17450101.2019.1601828.

SYKE (2020) *Suomi ja kestävä hyvinvointi*. Available at: https://www.syke.fi/fi-FI/Suomi_ja_kestava_hyvinvointi.

Tennberg, M., Emelyanova, A., Eriksen, H., Haapala, J., Hannukkala, A., J.K. Jaakkola, J., Jouttijärvi, T., Jylhä, K., Kauppi, S., Kietäväinen, A., Korhonen, H., Korhonen, M., Luomaranta, A., Magga, R., Mettiäinen, I., Näkkäläjärvi, K., Pilli-Sihvola, K., Rautio, A., Rautio, P., Silvo, K., Soppela, P., Turunen, M., Tuulentie, S. and Vihma, T. (2017) *The Barents area changes – How will Finland adapt? [Barentsin alue muuttuu – miten Suomi sopeutuu?]*. Helsinki: Valtioneuvoston kanslia (Valtioneuvoston selvitys- ja tutkimustoiminnanjulkaisusarja, nro 31/2017). Available at: http://pure.iiasa.ac.at/id/eprint/14469/1/The%20Barents%20area%20 changes.pdf.

Toivanen, R. (2003) 'Saami people and the Nordic civil societies', in Götz, N. and Heckmann, J. (eds.) *Civil society in the Baltic Sea Region*. Aldershot: Ashgate, pp. 205–216.

Toivanen, R. (2019) 'European fantasy of the Arctic region and the rise of Indigenous Sámi voices in the global arena', in Sellheim, N., Zaika, Y. V. and Kelman, I. (eds.) *Arctic triumph: Northern innovation and persistence*. New York: Springer International Publishing, pp. 23–40.

Toivanen, R. and Cambou, D. (forthcoming) 'Human rights', in Krieg, C. P. and Toivanen, R. (eds.) *Situating sustainability: A handbook of contexts and concepts*. Helsinki: Helsinki University Press.

Toivanen, R. and Fabritius, N. (2020) 'Arctic youth transcending notions of 'culture' and 'nature': emancipative discourses of place for cultural sustainability', *Current Opinion in Environmental Sustainability*, *43*, pp. 58–64. doi: 10.1016/j. cosust.2020.02.003.

Tsing, A. (2007) 'Indigenous voice', in de la Cadena, M. and Starn, O. (eds.) *Indigenous experience today*. London: Routledge, pp. 33–68.

Virtanen, P. K. (2012) *Indigenous youth in Brazilian Amazonia: Changing lived worlds*. London: Palgrave.

Wiegel, B., Boas, I. and Warner, J. (2019) 'A mobilities perspective on migration in the context of environmental change', *Wiley Interdisciplinary Reviews. Climate Change*, *10*(6). doi: 10.1002/wcc.610.

Wodak, R. (2015) 'Critical discourse studies: History, agenda, theory and methodology', in Wodak, R. and Meyer, M. (eds.) *Methods of critical discourse analysis*. 3rd edn. London: Sage, pp. 1–22.

Zickgraf, C. (2019) 'Keeping people in place: Political factors of (im)mobility and climate change'. *Social Sciences*, *8*(8), p. 228. 10.3390/socsci8080228.

Interviews

008 – Glocal_XX_8female_6.2013.
010 – Glocal_XX_10male_6.2013.
024 – Glocal_XX_24female_6.2013.
041 – D 40: 041_Glocal_XX_RYHMÄ_ATLAS
045 – D 44: 045_Glocal_XX_F_ATLAS

5 Youths' and their guardians' prospects of reindeer husbandry in Finland

Tanja Joona and Pigga Keskitalo[1]

Introduction

This chapter investigates what young reindeer herders think about their liveli-hood and future prospects to continue in this profession in Finnish reindeer herding districts. In addition, their guardians are interviewed so that the overall family perspective is available. Reindeer herding is a traditional liveli-hood practised across the Arctic and Subarctic regions, areas that have become the intense focus of development and policy interests. The Arctic area is predicted to experience dramatic climate shifts over the coming decades as a result of climate change. Young reindeer herders today face an uncertain future. Reindeer herders' pastures are being lost, and the full range of the consequences and impact of climate change is unclear. The conditions under which they practise herding today may be very different by the time they reach middle age. It is crucial for the future of reindeer husbandry that the youth adopt positive thinking, know how to assess opportunities and reach out to the future. The youths' choices are important when it comes to the future of reindeer herding—whether or not they decide to become rein-deer herders.

The study cases are in the following selected areas of reindeer herding cooperatives: Palojärvi in the town of Ylitornio (South-West of Lapland), Kaldooaivi in the municipality of Utsjoki (North of Lapland) and Näkkälä in the municipality of Enontekiö (North of Lapland) (see Reindeer Herders' Association 2020a, p. 11). According to previous studies, reindeer herding is being affected by the pressure of competing land use (Pogodaev and Oskal 2015), climate change impact (Vuojala-Magga et al. 2011; Box et al. 2019) and fear of an unstable future (Kaiser 2011; Omma et al. 2011; Kaiser et al. 2013, 2010). Mirroring a great deal of external stressors (Arctic Council Sustainable Development Working Group [SDWG] 2015), it is imperative to understand the rapid changes in the Arctic and ways to improve adaptation practices, what kind of views the youth hold and what kind of ideas they have with regard to the future of the reindeer herding and whether young people are willing to continue the livelihood. This chapter increases the knowledge on the future prospects of young reindeer herders in Finland. The knowledge can be used when developing the conditions of young

DOI: 10.4324/9781003110019-8

herders as well as the northern communities as a whole. Instead of following the global trend of immigration to cities, Indigenous communities can stay viable with strong cultural ties to traditional lands and natural resources (United Nations 2009, p. 93).

Reindeer herding as it exists today began in the late Middle Ages. The Sámi started herding reindeer and migrated nomadically with their herds according to the rhythm of the seasons. The Finnish population also adopted reindeer herding early on and developed it further for their own needs. Reindeer husbandry considers the characteristics of each area, such as the terrain, settlements and natural conditions (Reindeer Herders' Association 2020a, p. 17).

Finnish Lapland covers about a third (100,369 km^2) of Finland's total area. It is sparsely populated, the density being just two inhabitants per square kilometre, and the total population is 179,000 inhabitants. The area is known for its peculiar landscape, wilderness and traditional livelihoods, especially reindeer herding, which is an important part of the local economy. Due to the increase in the global prices of energy and raw materials, especially minerals, a rapidly expanding invasion of industrial land use and exploitation of natural resources is occurring in Lapland. Reindeer and people have a connection that is thousands of years old, first through deer hunting, then through reindeer domestication and herding. The profession of a reindeer herder is based on information and traditions that have been passed down through generations. In their work, nature and the eight seasons of the year are always present. Pure nature is very important for the continuity of herding work and reindeer herding culture. Both women and men can be reindeer herders (Reindeer Herders' Association 2020b). However, the position of women reindeer herders can be paradoxical because of changes in society, for example desire to get education and demands of traditional society (Kaiser et al. 2015).

Many of the older generations of reindeer herders remember the days when there were expansive grazing lands (Vuojala-Magga and Turunen 2015), the predator stock was kept under reasonable control (Wennstedt 2002; Tveraa et al. 2013), the reindeer herd was dense and the whole village gathered at the fences—both reindeer and the rest of the village and nearby villages (Heikkinen 2006). However, there are very viable communities still today (Pekkarinen 2006; Sarkki et al. 2016; Lépy et al. 2018). It is clear that the standard of living in Finland has increased over the generations. Finland has progressed from one of the poorest corners of Europe in the late nineteenth century to a highly advanced and innovative industrial country ranked by assorted international indices as the happiest, most stable, safest and best-governed country in the world (Reiter and Lutz 2019, p. 2; Statistics Finland 2019; Voutilainen 2016). According to Kokkinen (2012), considerable input into education has had a significant relation with the dramatically changed economic performance in Finland. The Finnish economy has experienced remarkable growth over the last 150 years, as far as statistics are available. According to Ojala et al. (2006), Finland was a fully agrarian economy that was heavily regulated and reliant on foreign trade. Even as late as the outbreak of the Second World War, the majority of the Finnish population was still employed in agriculture. It was only during the latter part of the twentieth century that Finland became a highly versatile economy, with both industries and services blossoming simultaneously. This progress was accompanied by the development of an extensive welfare state, high levels of investment in education and a considerable population shift from the countryside to the growing cities of southern Finland (Ojala et al. 2006; Reiter and Lutz 2019, p. 4). These societal and economic changes have also affected reindeer husbandry. In spite of Finland being one of the wealthiest countries in the world, Indigenous peoples' rights to lands and waters are still not recognized, which complicates land usage issues related to reindeer husbandry. Getting stuck in the past means paralysis in reindeer herding if only the past is seen as positive (Reinert et al. 2009). If instead of looking for opportunities and the positive impact of today's reindeer herding and future prospects we become paralysed by the fear of change, the world begins to look hopeless, and the future runs its course negatively affecting children and young people. When the world begins to look hopeless, young people often tend to seek something else. Reindeer husbandry—if nothing else—provides an excellent starting point for those willing to work as an entrepreneur and professional reindeer herder. The work is independent, close to the traditional living style of the local culture and conducted outside with animals. However, there are also many challenges—the profession is prone to accidents and vulnerable to the disadvantages of other forms of land use. Fortunately, the reindeer herder entrepreneurship field has been able to build on positive premises rather than negative—the share of reindeer owners under the age of 25 is higher than the share of retired people. Of course, traditionally, people of all ages play a valuable role in the practical work of reindeer herding. Children grow up to care for reindeer from their infancy, and the elderly participate in the work for as long as their health allows. The work is done together, taking turns caring for the others. Reindeer husbandry bridges generational gaps (Ollila 2019, 2014).

In Finland, there are around 900 reindeer owners under the age of 25, located in every part of the broad reindeer husbandry area from north Finland to the southern areas of the reindeer herding region (see Figure 5.1), and they are actively involved in reindeer herding. There are girls and boys of every age from the north, centre and south of the reindeer herding district (Jernsletten and Klokov 2002; Reindeer Herders' Association 2020c). The willingness of young reindeer herders to continue herding is obviously the principal element for continuity. The good news is that there are young people interested in becoming reindeer herders. Only the southern reindeer husbandry area is experiencing problems with this, mainly due to long-term problems with large carnivores (Reindeer Herders' Association 2020c).

Many youngsters make a very important decision for the future of reindeer herding—they decide to become reindeer herders. Every year, several dozen youths decide to take a reindeer husbandry entrepreneurship education at the Sámi Education Institute in Inari, Finland, to qualify for start-up aid in the

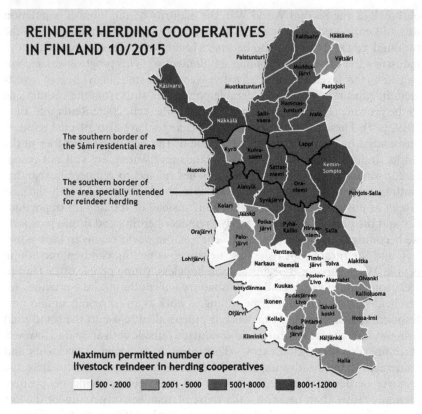

Figure 5.1 Reindeer Herding Districts in Finland and the numbers of reindeer.

(Source: Reindeer Herders' Association 2015)

profession (Sámi Education Institute 2020). The institute not only teaches how to become a professional reindeer herder; it also teaches how to utilize reindeer meat and skin, how to make handicrafts and clothes from reindeer horns and leather, for example. It is expected that to be able to work as a professional reindeer herder with economical support, one needs education in the reindeer herding entrepreneur profession and needs to have a minimum requirement of 80 living reindeer. In Finland, reindeer herding can be studied as a vocational education at the Sámi Education Institute Toivoniemi campus in Kaamanen, a village 30 kilometres north of Inari within the precincts of suitable buildings and structures. There are three kinds of study options. The first is a full basic study programme that consists of two to three years of study conducted on the campus and reindeer herding farms. The second programme is learning to work as part of an apprenticeship training and is conducted mainly at their own family reindeer herd or related places. Students go to campus to study and meet the teachers, and then they do their skills demonstration. The third option is open vocational education training that gives study credits (Frangou and Keskitalo 2020; Keskitalo et al. 2020).

The lives of young reindeer herders are governed differently than in the days before the societal changes, changes in profession, technical development and aspects affecting the profession (Reindeer Herders' Association 2020b). Times change, and life changes with it. Livelihoods and cultures, including reindeer husbandry, are also changing, and reindeer herders adapt with the changes (Rasmus et al. 2016).

We consider people under 29 years of age young reindeer herders, according to Finland's Youth Act, which defines youth to be those under 29 (Nuorisolaki 1285/2016). In the opinion of young people, the definition in the Youth Act is too broad. According to the young people themselves, youth begins at around the age of 10 and ends around 20 (Merikivi, Myllyniemi and Salasuo 2016).

The interviews aim to answer the overall research question of the youth's connectedness to reindeer herding. The sub-questions are as follows:

1. What are the first memories of reindeer herding in the youths' lives?
2. How is the family occupied in the reindeer herding livelihood?
3. What prospects do the youths perceive in future reindeer herding?

The aim of the research question is to discover the overall ideas that the youth have about reindeer herding through their narratives and their prospects in the future to open their minds to future ideas about their career choices. We have asked questions to illuminate the youth's connectedness to reindeer herding.[2] The thematic interview contained background questions concerning the ages and genders of the interviewees (N = 12). The interviews were semi-structured, and a data-driven content analysis was performed (Anderson 2004).

Studies about Indigenous youths' prospects in the Artic context

This chapter is interested in processing the ideas of young reindeer herders and their guardians about the entrepreneur in the Arctic context in Finland. In the literature review, we looked at what other researchers have stated about reindeer herding as a profession and, more specifically, what the youth think according to previous studies about the profession to explain future prospects.

Most remote communities in the Arctic share their fate with Europe's northern periphery in that demographic development is characterized by a population decline that has been going on for decades (Karlsdottir and Jungsberg 2015). According to recent and former studies (e.g. Heleniak 2020; Nymand Larsen et al. 2010), it is often difficult to get skilled young people that have moved away because of studies, work and other reasons to come back to their home regions to help create growth and optimism. One explanation for this problem is that there is no variety of jobs or education possibilities (Keskitalo 2019). In many cases, there is also a lack of women because of limited job opportunities and because they consider the opportunities for personal expression and recognition insufficient. Many young people across the Arctic region imagine themselves holding secure, well-paid jobs in the future, and they see education as one way to achieve this. Some young students from the Faroese Islands argue that future education will be even more ambitious than it is today if they are to catch up with technological developments in the next 20 years or so. Students are even ready to do extracurricular activities for this reason. The idea that future young people have to learn much more than today indicates that learning is seen as a means of societal development (Karlsdottir and Jungsberg 2015). Reindeer husbandry, however, kept villages populated, so it had a strong regional impact (Luke 2016).

When looking at the mobility of young people, living in the countryside can be seen as a choice in terms of the development of urbanization—and for a growing number of people, a wider lifestyle choice. The questions are: who stays, and who leaves? For those families who have stayed, the adults had to make a choice. Young people choose what to do after primary school, where to go to study their vocational education and then what to do after. The people living in the countryside do not necessarily expect urban services, but the place of residence is chosen according to what is considered important in life. The constant striving for material gain has been questioned. The significance of leisure time, human relationships and physical exercise in contributing to people's happiness addresses questions of the extent to which the countryside has a kind of 'wellbeing surplus' in the minds of the young (Nieminen-Sundell 2011). The findings are blended. Youth are at a stage in their lives where they want to travel, and many of them will not settle as established grown-ups for 10 to 15 years. In many ways, the youth period in life has become extended with increased formal requirements on training and education until one enters life as a fully acknowledged adult citizen (Evans, 2018). This state of transience is in many ways liberating, but it can also feel troubling. Looking at the youth from the Arctic regions, they are attracted to more urban settlements in the short term, whereas in the 20–25-year perspective, it could be interesting to live in the area of their upbringing. One dominating trend is how the youth's lifestyle is, to a large extent, connected to an urban settlement. Therefore, it is difficult for the youth to 'realize' themselves as young people staying in a rural area in the short term. This is also reflected in their ambitions of being mobile to pursue education and work (Jones 2002; Hoolachan et al. 2017). Similarities can be found all over the Arctic.

According to Kaiser et al. (2013), the experience of the young reindeer herders was that being a reindeer herder is a privileged position which also implies many impossibilities and unjust adversities they have no control over and that there is nothing they can do but 'bite the bullet' or be a failure. The Eallin—the reindeer-herding youth project conducted by the Arctic Council SDWG (2015)—gives good examples of how the future of reindeer herding is seen by Sámi youth from Fennoscandia, Nenets and Khanty in the Yamal-Nenets Autonomous Okrug; Dolgan, Chukchi, Yukagir, Even and Evenki in the Republic of Sakha (Yakutia), Russia; Evenki from China; and the herding youth from Mongolia. To sum up, the Eallin project showed that the youth are greatly concerned about the loss of grazing land. The loss and disturbance of reindeer pastures are primarily related to mining, windmill extraction and infrastructure development. Predators and protected areas are also major concerns. Encroachments have had a large impact on reindeer husbandry in many places; thus, the reindeer-herding youth are worried that this could be a considerable threat to their future (International Centre of Reindeer Herding 2015; Uboni et al. 2016). Gradual encroachment combined with climate change is challenging the traditional livelihood of herders. Reindeer herders wish to be heard at an earlier stage in the planning processes of industry as well as contribute knowledge with regard to changes that will impact their livelihoods (International Centre of Reindeer Herding 2015).

One of the consequences of climate change, globalization and changes in lifestyle is the emergence of a situation where more females consider migrating permanently away from their home community and region, and, indeed, increasing numbers of young women actually do so. Of course, gender-based differences in migration choice are nothing new in the Arctic. In connection with large-scale resource development projects, young and middle-aged males in search of employment and income opportunities have chosen to become migrant workers, leaving their communities for either a shorter or a longer period of time in the process. Seldom have they left the community permanently. Only if the job turned out to be more permanent in character and generated substantial income have they done so. In such cases, they often arrange with their families to follow them and settle in the new town or village. Females, however, seem to migrate more permanently. Moreover, such choices have significant implications for the communities they leave, for instance, decreasing opportunities for marriage and the maintenance of family life and family structures as well as fundamentally influencing other cultural activities (Rasmussen 2009). The gender-related perception of customary male activities related to resource exploitation seems to be 'sticky' in the sense that males have difficulty in moving on from what once were key activities but now constitute only a small percentage of the available jobs. Females, however, seem less limited by specific job characteristics, determined by what may be considered 'traditional' and 'acceptable' activities.

Another interesting study on Nordic Arctic youth and their future perspectives in connection with the project *Foresight Analysis for Sustainable Regional Development in the Nordic Arctic* was commissioned by the Nordic Working Group for Sustainable Regional Development in the Arctic (Karlsdottir 2015; Karlsdottir and Jungsberg 2015). The study investigates social sustainability involving questions about attracting and/or keeping young people, especially women, in peripheral communities. This is a fundamental issue in keeping these communities viable, inhabited and attractive for everybody. The study involved several topics in which we are especially interested— culture and Indigenous traditions as well as the impact of social media on youth. The study shows how social media platforms can be used for empowerment purposes in policy issues concerning land use. Social media also provides connections to non-Sámi youth who are supporting the work of ensuring Sámi people acquire rights and recognition. This also indicates how young Sámi today are a mixed group where cooperation is less about ethnicity and instead based on shared values and future visions (Karlsdottir 2015; Karlsdottir and Jungsberg 2015; Laitala and Puuronen 2015; Öhman 2015).

Some general guidelines mark the future of their traditional livelihood, reindeer herding. One study suggests that youth should be taken into entrepreneurship in the early stages (Fleming et al. 2015). Other studies also indicate the importance of state-supported training and resourcing for Indigenous needs (Chantrill 1998; Miller 2005; Barber and Jackson 2017). Individuals need motivation and support from their family, friends and community to adapt to the work culture and successfully get and keep a job (Haley and Fisher 2014). Their early memories partly explain the connectedness and family expectations and roles. According to studies, the family has an important role in the background of the youth's expressions in addition to personal motivation, gender, friends and family background affecting the career choices of the youth (Super et al. 1996; Chen 1997; Bandura et al. 2001; Ferry 2006). For example, youth coming from agricultural backgrounds most probably also choose to work in agriculture (Fizer 2013). The place, cultural background and time shape youths' identities and career choices (Arbona 2000).

Moreover, the kinds of measures taken towards the youth in rural and Indigenous communities are of utmost importance (Indigenous Peoples Forum at IFAD 2017). The sense of cultural and social belonging seems to be explained by means of working with traditional livelihoods, which has a further impact on the employment of young people in remote, sparsely populated Arctic regions (Veijola and Strauss-Mazzullo 2018). According to the research, youths have identified and reported a shared problem of trying to be successful in the face of the sometimes contradictory demands of their Indigenous culture and those of the dominant one (Mbunda 1983; United Nations 2009; Gillan et al. 2017). For example, many young people feel confronted with a hard choice between continuing school or staying in their community if there are no further education possibilities after primary school (West et al. 2010). Research has shown that this implicates a cluster of problems, including lack of jobs in their local communities, outmigration to find employment, the unavailability of local housing, high living costs in remote rural villages and leaving their families as well as abandoning their aspirations (Ulturgasheva 2012a, 2012b, 2014, 2015; Ulturgasheva et al. 2014).

The vulnerable situation of reindeer herders and the impact of climate change may have serious consequences for trade and herders' overall way of life (Furberg et al. 2011). Reinert et al. (2009) claimed that the key to handling permanent changes successfully is that the herders themselves have sufficient degrees of freedom to act. Further, research shows that the most important strategy of reindeer herders is constant adaptation to changing conditions (Reinert et al. 2009).

According to statistics, the number of reindeer herders in Finland has shrunk over the last 20 years; the number of men decreased from 6,000 to 3,000 between 1990 and 2018, whereas the number of women has remained roughly unchanged (Figure 5.3). Reindeer herding can be different

Figure 5.2 Youths during a break from summer reindeer calf marking.

(Photo: Tanja Joona)

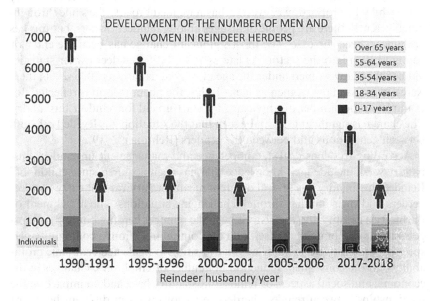

Figure 5.3 The numbers of men and women in reindeer herding, 1990–2018.

(Source: Reindeer Herders' Association 2019)

depending on the area in which one lives. Youths tend to participate in family reindeer herding work from childhood. Most youths get their reindeer mark[3] when they are small, and they own reindeer in addition to everyone in their family. The living style of reindeer herding is learnt through participating in the workings of a herding and experiencing its fluctuations over time.

Reindeer are semi-wild animals and require care for part of the year by a reindeer herder. In the northern part of Finland's reindeer herding area, reindeer graze more freely, and they require constant herding by reindeer herders. During the winter, many reindeer herders feed them even if the reindeer are not in fences because the pastures are becoming smaller depending on the area. In summer or autumn, there are reindeer markings. Some reindeer herders make hay for reindeer. Reindeer herders repair fences and maintain equipment. In autumn, there are reindeer round-ups and all the work involved both before and after that. Further, the work is carried out in various kinds of weather conditions (Joona 2018; Pirttilä 2020).

According to Rehtonen (2019), there are two sides to the development of the age structure of reindeer owners. There are increasingly more young people working as reindeer herders—especially young women. Until 2014, the upper age limit for individual aid was 65 years. However, under EU law, the restriction was considered age discrimination. As a result of the abolition of the age limit, the share of retired people, mainly men, in reindeer herders has increased by 150 since 2014. The Reindeer Herders' Association is concerned about this development since younger people cannot continue working if their parents do not give up the work. Those leaving the sector have been predominantly male, while the proportion of women has remained broadly the same throughout. Currently, there are people of all ages among reindeer owners, but this has not always been the case. The most significant change has been the entry of young women into the sector. As late as 1995–1996, reindeer owner statistics did not show any women under the age of 17, but as early as 2005–2006, there were almost as many women as men. Today, the phenomenon is reflected, for example, in the number of applicants for start-up aid for reindeer husbandry. The change is significant. It can be seen that the situation has levelled off both between generations and between the genders (Rehtonen 2019).

According to Oinas (2018), competence and management in reindeer husbandry work include tacit knowledge that has been acquired since childhood. Reindeer husbandry has several different means of survival that carry reindeer herders and their families through challenging situations. The attachment of the lifestyle identity and the historicity make reindeer herding binding on most generations and binding on itself and the community. From a community perspective, the lifestyle is in many ways a transition to weakened economic profitability and the consequent reduction in number of employees. Changes in the economic and social aspects of reindeer husbandry have had an impact on the relationships between reindeer herders in various ways in different herds. At one extreme, the spaces may have tightened and escalated, while those left at the other extreme have condensed further cooperation, and the importance and contribution of everyone involved is highlighted. The decisive factor for the future of an individual group is whether the situation of the group is such that, according to new, young generations of reindeer herders, entry into reindeer husbandry work is still sensible and possible (Oinas 2018).

According to Daerga et al. (2008), men and women from reindeer-herding families need partly different conditions to enjoy a high quality of life. From their results, it might be predicted that poor somatic and psychosocial health, increased intrusion from exploiters on the grazing land and declining profit in reindeer husbandry constitute important threats to a good quality of life among members of reindeer-herding families (Daerga et al. 2008). According to Buchanan et al. (2016), rural communities that rely on natural resources around the world demonstrate a highly gendered division of labour, and pastoralists such as reindeer-herding communities appear to be no exception. Young and older female and male reindeer herders reported that there are different expectations for men and women in households regarding who should support reindeer husbandry and how this should be done. Boys are frequently socialized into the profession differently than girls (Buchanan et al. 2016).

The voices and views of the young herders and their guardians

Connectedness and memories about reindeer herding

The connectedness of reindeer herding tells about sociological processes of an Indigenous traditional living style (Balto 1997). The first memories related to reindeer herding are gained in early childhood. The socialization process into the livelihood is done through the family and the living style of the family. All the interviewees have had reindeer in their families. We were interviewing those actively occupied in the profession.

> The first memories are when I was three or four years old from calf marking.
>
> (young male)

All the interviewees are from families who have been reindeer herders for generations, and the living style has been passed down from generation to generation, from grandparents to sons and their children or near kin.

When asked who is the most active in the reindeer-herding family, the answers varied. Often youths answer that it is the family's father. It seems that most of the active reindeer herders are men, according to the interviews, so the profession is often gender-tight.

> Most active reindeer herder in my nuclear family is my father.
>
> (young female)

Or sometimes the extended family:

> My uncles and cousins have an active reindeer herding activity.
>
> (young male)

Still, there are also women in the livelihood. According to Joks (2005), women serve an invisible but important part of reindeer herding, doing a lot of things that are not so obvious, for example, taking care of things at home when the men are herding in the forest or mountains. Mothers or female relatives also look after the young children during the actual reindeer herding works, such as calf marking in the summer or separation and slaughtering in the fall.

The home means a lot for reindeer-herding youth. An interviewee claims that there is a willingness to come back after studies, meaning that one must often move away from home to study. The youths' ideas about studying elsewhere consist of traveling and getting new insights into life:

> I want to live in the same place in a home village in the future. Before that, I'm going to study elsewhere, I still don't know where. Possibly, I will also travel during these years.
>
> (young female)

The youths consider their connectedness to their livelihood as part of their everyday life. Herding and its related activities are part of the annual rhythm still practised strongly in rural Lappish areas. It can be considered that reindeer herding is the centre of life and that all other things happen on its terms. If there are schooldays in the middle of the slaughtering season, kids usually stay at home and participate in the reindeer separation.

Nowadays, reindeer herding is a blend of traditional and modern life and technology. Reindeer herding is most often passed down in a family, but there are different ways to become a reindeer herder, such as through kin, as a family heritage or through learning the tradition in other ways. All youths need to receive an education to become professional reindeer herders and so they can raise grants. Of course, it is possible to be a reindeer herder without becoming a professional and instead engage it as a second occupation. It seems as though reindeer husbandry is a gender-tight and male-oriented occupation. Still, there are now increasing numbers of women and young girls who wish to become reindeer herders. They can practise differently according to the principles of the surroundings, sometimes taking part more actively in the work that men do or by doing different kinds of tasks in their home yards if there are children around. It is still a fact that reindeer herding is physically demanding work and prone to accidents because of the demanding context in working with animals and the demands from the natural surroundings. The youths explain that the living style with different kinds of seasonal work and living with reindeer herding in their home village is motivating. They are ready to see the profession as a part of their future life, either as a main profession or second occupation.

Ideas about reindeer herding

The youth think that reindeer herding is an important vocational branch, although there are things that might harm the living style, such as climate change. Reindeer husbandry is considered an industry that is vulnerable to the negative effects of climate change. Changing weather and snow conditions especially impact the reindeers' ability to find food (Kumpula 2012).

Fortunately, reindeer herders have access to technologies and treatments that have already helped the reindeer herding livelihood adapt to the changing conditions. Reindeer husbandry does, however, still have several challenges to overcome in the process of adapting to climate change, but it is also a livelihood with a relatively high adaptation potential to the changes (Rasmus et al. 2020).

> Climate change is a nasty thing. It is likely to affect the food intake of reindeer as we have already been able to realize.
>
> (young male)

The social nature of reindeer herding is, in many ways, the driving force for young people. Summertime with calf markings is a learning time for young people. The different parts of a round-up enclosure become familiar for kids. Reindeer are handled in the churn (*kirnu* (Fin.) *girdno* (North Sámi)) when it is slaughter time.

Youths participate in different kinds of tasks during the reindeer-herding year. The high season of reindeer herding is the most interesting period according to one youth:

> Reindeer calf marking and reindeer round-up are the most fun events during the reindeer year.
>
> (young female)

After all, it is fun for the youth when they get to spend the nights in a gang.

> From learning to like, it goes into the blood. Besides, they really help a lot.
>
> (parent of an interviewee)

The youths share a realistic picture of reindeer herding because they are socialized into it from childhood and participate in different kinds of tasks. They learn that the work is physically demanding, but many still want to wander with the herdsmen tens of kilometres during the day when the reindeer are gathered for separation and slaughtering in the fall. In the summer, the work takes place normally at night-time. For example, one 16-year-old male interviewee said he started at the age of 14 and walked 25 kilometres during his first night in the forest. The youth also find the different kinds of tasks interesting, and the high season of reindeer herding is a much-appreciated occasion. The youths' participation in entrepreneurial work is a natural part of family life. Everyone has their own role, depending on the age and task. Everything happens often with the support of the guardians.

Future prospects

There are many kinds of ideas about the future among the youth. A young primary school-aged girl claims that she is about to study another profession, but also wants to work in reindeer husbandry:

> I am planning to study a profession but also probably continue reindeer husbandry. I will go to high school after elementary school, after which I plan to study more. I don't know yet what profession I'm studying for. I plan to continue caring for reindeer alongside the work. I'm not going to start to work as a full-time reindeer herder.
>
> (young female)

It seems that the girls are more willing to find a second occupation besides reindeer herding, which is natural as reindeer herding is a physical profession and is also male-oriented (Buchanan et al. 2016). It is also interesting that young females find reindeer herding very attractive and that they want to be involved with the profession in one way or another.Some youths think that the future of reindeer herding might be in crisis as they believe that the number of reindeer herders will be reduced further. The youths see that if reindeer husbandry is an ongoing profession in a family, the traditional living style is in good hands and also has a chance to be delivered to future generations (see also Oinas 2018).

> The practice of reindeer husbandry will decrease in the future. I'm not sure. Depends on future generations. In our family, reindeer husbandry is likely to continue as there are so many of the family involved in a profession and living style.
>
> (young male)

Many parents encourage children to continue in the reindeer-herding business, but they also highlight the importance of studying. This can be contradictory and requires balancing. It also reflects the intergenerational thinking and way of life.

> After elementary school, I'm going to go to high school and after that maybe apply for a reindeer husbandry line in Inari. Currently, the enthusiasm for reindeer herding is fierce, but let's see then.
>
> (young female)

> I am kind of unsure yet. I am planning what to do. I have to go to the army first and decide then. I, in a way, want to study in Inari, but I am unsure if I need to study something else.
>
> (young male)

The youths share contradictory ideas about reindeer herding. Youth is a period of insecurity and making choices. They are not quite sure what will happen in the future. Some of the interviewees are more dedicated to reindeer herding, while others are unsure and are planning to study something else. What is still certain is that everyone is willing to continue reindeer herding in some way, either as a main livelihood or as a second occupation paired with other earnings. What is also evident is that that each youth shares positive ideas about reindeer herding and that there is willingness to continue the livelihood. Alternatively, the modern way of life keeps the youths under pressure with obligations to the family and history and temptations typical for this age.

Reindeer-herding identity

Reindeer herders have a strong identity connected to practical work with reindeers. Identity forms at the early stages of childhood by learning from and imitating adults.

> I believe reindeer husbandry will generate income for me in the future. I have my own reindeer mark and my own reindeer. I think reindeer husbandry is a nice job and kind of a nice hobby or living style if not a main profession. I can recommend it to others.
>
> (young male)

Young people can already see the cultural connection with reindeer herding. It is not just a profession—it is a way of life.

> Reindeer herding can also be really different depending on which area of Lapland you live in. Everyone has their own challenges and goals, and that's why there's as many ways to herd reindeer as there are reindeer herders. Overall, reindeer herding is more than just a profession. It's a lifestyle. I am lucky to have grown up in this community and way of life, as it will always be a part of me in one way or another.
>
> (young female)

Figure 5.4 People of all ages play a valuable role in the practical work of reindeer herding.

(Photo: Tanja Joona)

Figure 5.5 Children learn by imitating, but possess also a strong curiosity by nature.
(Photo: Tanja Joona)

Figure 5.6 Young reindeer herder.
(Photo: Pekka Keskitalo)

Reindeer herding has been traditionally seen as a very male-oriented occupation. This is the case in many Arctic countries. In Russia, Indigenous women and men often express their concern about an imbalanced gender relation. The increasing absence of women in the taiga and tundra is considered one of the principal reasons for the crisis in the hunting and reindeer economies (Schindler, 1997; Ssorin-Chaikov, 2003). In addition, Indigenous men, particularly in reindeer encampments, complain about the loneliness of the single life and the lack of women to share household duties (Povoroznyuk et al. 2010).

Currently in Finland, there are reindeer owners of all ages, but this has not always been the case. The most significant change has been the entry of young women into the sector. As late as 1995–1996, reindeer owner statistics did not show any women under the age of 17, but by 2005–2006, there were almost as many women as men. Today, the phenomenon is reflected, for example, in the number of applicants for start-up aid for reindeer husbandry. According to the parents interviewed, the change is significant:

> This is a big thing that has hardly been able to outline before. It can be seen that the situation has levelled off, both between generations and between the genders.
>
> (parent of an interviewee)

It is clear that some kind of cultural change took place among reindeer herders in the early 2000s as an increasing number of girls were listed as reindeer owners. It appears that the profession has a high status, and, even for women, it is possible to study and become reindeer herders. The profession allows freedom in nature, and one can spend a lot of time outside and with animals, so the field is appealing. Reindeer herders are quite independent, and they can determine a lot by themselves and combine it with other income. Since one can study reindeer herding in a vocational education school, the profession can be seen as a real and potential profession for the youth. Still, many think that you must have some kind of connection to reindeer herding before it makes sense.

> The change of ownership of reindeer is mainly between families. Maybe the mind-set has changed. One would think that the social change in gender roles is also reflected here—reindeer husbandry is not separate from the rest of the world. The debate related to gender equality has reached here as well, and it no longer matters whether the successor is a girl or a boy.
>
> (parent of an interviewee)

The parents/guardians have a strong background position. One guardian stated that the parents support the youths in different ways but let them finally decide.

> Of course, we hope that many children in a family will choose to stay in a livelihood. We try to add them to the yearly and seasonal work but they

can decide by themselves whether they want to be more active or not. But we try to motivate them and give them a chance. I think it would be nice if youth could stay in a livelihood. But, finally, they have to choose by themselves. I think both girls and boys could be reindeer herders but I can understand as reindeer herding is a very physical profession that girls tend to educate themselves and then combine reindeer herding to the other professions or entrepreneurs. Of course, I am afraid if the youth will cope as the livelihood is quite tough. You have to have good talent and motivation to cope in a branch and you have to have a good starting point from the family, basically since childhood. I am happy to see that youth are interested in a livelihood—even in a combined form. It gives a lot of hope for the future of survival of the livelihood. I think reindeer herding has a good status nowadays and it can give a quite stable income if you work hard and plan your economy well.

(parent of an interviewee)

Conclusion

The aim of this chapter was to construct reindeer-herding youths' views about the livelihood and the future. This scholarly chapter is based on an analysis of quotes from young herders and their guardians. The picture of youth is filled with insecurity about the future, what to study and what to become later in life. In that sense, young reindeer-herding family members share a general picture of youth. Still, some are more committed to reindeer herding and secure than others. Finally, there is a shared feeling of excitement, courage and will, the base for the continuation of reindeer herding, as well as the other similar activities, including fishing, hunting and making handicrafts. The youths seem to not yet share the challenges in Finland as the male reindeer herders in Sweden (Kaiser et al. 2013). There are differences in legislation in the two countries, which pressure the reindeer herders differently. The views remain heterogeneous. In general, the views are optimistic while the different choices for life remain open. Still, all the youths saw reindeer herding as a richness and respected the contents and aspects of it.

While resource exploitation is still viewed as the main economic basis for communities in the North, the reality is that the third sector—the service sector with wage work in administration, education, social services, etc.—has become the main income source for most families. With limited job opportunities for well-educated women, however, the prospect of staying remains highly unattractive, resulting in the continuing outmigration as the only option available to them.

For many men, however, limited options exist in respect to leaving their current occupations. In the small villages of the north in particular, the situation is often desperate. Without proper qualifications, unskilled jobs become the only option, and as these are also now disappearing, the prognosis for unskilled male employment is becoming bleaker. Their incomes from traditional activities are not enough to enable them to profitably continue to work as these limited incomes do not enable them to re-invest in new equipment to expand their activities.

They also have difficulty finding young girls who are interested in staying in the villages, thereby severely limiting the option of generating double incomes, which are needed to maintain a life entailing traditional activities. Without skills and money, it is not possible to move to larger places to find work. This state of affairs then sees many in desperate situations accompanied by violence and abuse, which only adds to the female flight to pursue a better future and to avoid the negative consequences of the process of decline.

The villages are the first to be abandoned, though the smaller towns are now also suffering from female flight. Only the towns with higher education opportunities and a broader supply of qualified job opportunities seem to be able to maintain an environment attractive enough for younger women to enjoy. This is something very alarming and requires attention from decision-makers.

However, it is considered that the youth already in the early stages have ideas of choosing to be a professional reindeer herder, and this seems to be a gender-specific issue (Povoroznyuk et al. 2010; Vitebsky 2010). Males have stronger opinions about staying in the vocational branch of their families or kin. Even if girls decide not to have reindeer husbandry as their main profession, they still see the value of working and staying in reindeer herding— meaning to have another career but also work with reindeer. Reindeer herding has strong economic and regional value as families stay in the villages because of reindeer herding. Girls are becoming more aware that they can become reindeer herders, even as a second occupation. This tension explains the change in the roles of women: young reindeer herding women stand out in a positive way. It looks like the times have changed, and women have started to see the possibilities of living in traditional livelihoods to a wider extent. After the Second World War, more people, and, in particular, women, migrated from Finland to Sweden and to bigger cities from northern villages although there were and are still women who wish to stay in the livelihood and villages.

It seems that it is possible to continue living in a traditional way. Reindeer have always been and remain the foundation of reindeer-herding peoples' lives. Reindeer provide people with shelter, food, clothing and security and are at the centre of herding peoples' universe, the foundation of their cultures, languages, worldviews and ways of knowing. Reindeer are also the foundation of the reindeer-herders' economy (Lindqvist 2009; Reinert et al. 2009).

The reindeer herders themselves are the ones caring for the future of their culture, and society should create the conditions for this. Youths are the future of reindeer-herding peoples everywhere. The option to be socialized into reindeer herding needs to be a valued thing in the primary-school-aged youth in reindeer herding districts. School authorities and curricula need to consider reindeer-herding values and the youths in reindeer-herding families. Ideally, the local curriculum would give youths every kind of knowledge so that they can cope whether they choose to leave or stay. Local and more general knowledge is needed. More debate is needed in society regarding northern village life and policies to support people to live their chosen way.

Acknowledgements

We sincerely thank the youths and their guardians who were interviewed for this article.

Notes

1 The first author is a researcher with decades of experience in researching Indigenous and human rights contexts. The second author similarly has decades of experience in multicultural contexts and extensive experience with educational and intercultural research. Both authors have a strong connection to reindeer husbandry as both authors are integrated in reindeer herding and are also mothers to four children.
2 See the interview questions at the end of this chapter.
3 Each reindeer owner has one's own earmark. According to the Reindeer Herders' Association (2020a), there are 21 different markings known as 'deeds' and around 12,000 earmarks currently in use.

References

Anderson, R. (2004) 'Intuitive inquiry: an epistemology of the heart for scientific inquiry', *The Humanistic Psychologist, 32*(4), pp. 307–341. doi: 10.1080/08873267.2004.9961758.

Arbona, C. (2000) 'The development of academic achievement in school aged children: precursors to career development', in Brown, S. D. and Lent, R. W. (eds.) *Handbook of counseling psychology,* 3rd ed. New York, NY: Wiley, pp. 270–309.

Arctic Council Sustainable Development Working Group. (2015) *Youth the future of the reindeer herding peoples: Executive summary.* Available at: https://oaarchive.arctic-council.org/bitstream/handle/11374/1477/SDWG_EALLINN_Doc2_EALLINN_Executive_Summary_AC_SAO_CA04.pdf?sequence=1&isAllowed=y (Accessed: February 18 2021).

Balto, A. (1997) *Samisk barneoppdragelse i endring.* Oslo: Ad Notam Gyldendal.

Bandura, A., Barbaranelli, C., Caprara, G. V. and Pastorelli, C. (2001) 'Self-efficacy beliefs as shapers of children's aspirations and career trajectories', *Child Development, 72*, pp. 187–206. doi: 10.1111/1467-8624.00273.

Barber, M. and Jackson, S. (2017) 'Identifying and categorizing co-benefits in state-supported Australian indigenous environmental management programs: international research implications', *Ecology and Society, 22*(2). Available at: http://www.jstor.org/stable/26270133 (Accessed: February 19 2021).

Box, J. E., Colgan, W. T., Christensen, T. R., Schmidt, N. M., Lund, M., Parmentier, F. J. W., ... & Olsen, M. S. (2019). Key indicators of Arctic climate change: 1971–2017. *Environmental Research Letters, 14*(4), 045010.

Buchanan, A., Reed, M. G. and Lidestav, G. (2016) 'What's counted as a reindeer herder?Gender and the adaptive capacity of Sami reindeer herding communities in Sweden', *Ambio, 45*, pp. 352–362. doi: 10.1007/s13280-016-0834-1.

Chantrill, P. (1998) 'Community justice in Indigenous communities in Queensland: prospects for keeping young people out of detention', *Australian Indigenous Law Reporter, 3*(2). Available at: https://heinonline.org/HOL/LandingPage?handle=hein.journals/austindlr3&div=20&id=&page (Accessed: Feb 19 2021).

Chen, C. P. (1997) 'Career projection: narrative in context', *Journal of Vocational Behavior, 54*, pp. 279–295. doi: 10.1080/13636829700200012.

Daerga, L., Edin-Liljegren, A. and Sjölander, P. (2008) 'Quality of life in relation to physical, psychosocial and socioeconomic conditions among reindeer-herding Sami', *International Journal of Circumpolar Health, 67*(1), pp. 10–28. doi: 10.3402/ijch.v67i1.18223.

Evans, K. M. (2018/1998) *Shaping futures: learning for competence and citizenship.* New York, NY: Routledge.

Ferry, N. M. (2006) 'Factors influencing career choices of adolescents and young adults in rural Pennsylvania', *Journal of Extension, 44*(3), pp. 205–212. doi: 10.5367/ihe.2015.0253.

Fizer, D. (2013) *Factors affecting career choices of college students enrolled in agriculture.* Master's thesis, University of Tennessee, Martin. Available at: https://www.utm.edu/departments/msanr/_pdfs/Fizer_Research_Project_Final.pdf (Accessed: 18 February 2021).

Fleming, A. E., Petheram, L., and Stacey, N. (2015) 'Australian Indigenous women's seafood harvesting practices and prospects for integrating aquaculture', *Journal of Enterprising Communities: People and Places in the Global Economy, 9*(2), pp. 156–181. doi: 10.1108/JEC-08-2014-0013.

Frangou, S.-M. and Keskitalo, P. (2020) *Substantiating vocational competency identity in reindeer herding studies.* Master's thesis, Oulu University of Applied Sciences. Available at: https://www.theseus.fi/handle/10024/340624 (Accessed: 18 February 2019).

Furberg, M., Evengård, B. and Nilsson, M. (2011) 'Facing the limit of resilience: Perceptions of climate change among reindeer herding Sami in Sweden', *Global Health Action, 4*(1), p. 8417. doi: 10.3402/gha.v4i0.8417.

Gillan, K., Mellor, S., and Krakouer, J. (2017) *The case for urgency: Advocating for Indigenous voice in education.* Camberwell, Victoria: Australian Council for Educational Research. Available at: https://research.acer.edu.au/cgi/viewcontent.cgi?article=1027&context=aer (Accessed: February 18 2021).

Haley, S. and Fisher, D. (2014) 'Indigenous employment, training and retention: Successes and challenge in Red Dog Mine', in Gilberthorpe, E. and Hilson, G. (eds.) *Natural resource extraction and Indigenous livelihoods: development and challenges in the era of globalization.* London: Routledge, pp. 11–35.

Heikkinen, H. (2006) 'Neo-entrepreneurship as an adaptation model of reindeer herding in Finland', *Nomadic Peoples, 10*(2), pp. 187–208. doi: 10.3167/np.2006.100211.

Heleniak, T. (2020) 'The future of the Arctic populations', *Polar Geography.* doi: 10.1080/1088937X.2019.1707316.

Hoolachan, J., McKee, K., Moore, T., and Soaita, A. M. (2017) "Generation rent" and the ability to "settle down": Economic and geographical variation in young people's housing transitions', *Journal of Youth Studies, 20*(1), pp. 63–78. doi: 10.1080/13676261.2016.1184241.

Indigenous Peoples Forum at IFAD. (2017) *Synthesis of deliberations.* Available at: https://www.ifad.org/documents/36783902/40298813/Synthesis+of+Deliberations_

Third+Global+Meeting+Indigenous+Peoples+Forum+at+IFAD.pdf/9b62de47-b5f6-4aaa-9e0f-40df3a90b338 (Accessed 17 February 2021).

International Centre of Reindeer Herding. (2015) *Eallin: reindeer herding youth full report*. Kautokeino. Available at: https://www.scribd.com/document/262802467/Eallin-Reindeer-Herding-Youth-Full-Report (Accessed: February 18 2021).

Jernsletten, J.-L. L. and Klokov, K. (2002) *Sustainable reindeer husbandry*. University of Tromsø. Centre for Saami Studies. Available at: http://www.reindeer-husbandry.uit.no/online/Final_Report/final_report.pdf (Accessed: February 17 2021).

Joks, S. (2005) 'Boazosámi nissonolbmot - oaidnemeahttun geađgejuolgi', *Sámi diedalaš áigečála*, *1*, pp. 39–56.

Jones, G. (2002) *The youth divide: diverging paths to adulthood*. Bristol: Joseph Rowntree Foundation. Available at: http://www.bristol.ac.uk/poverty/ESRCJSPS/downloads/research/uk/3%20UK-Poverty,%20Inequality%20and%20Social%20Exclusion%20(the%20Youth)/Book%20(UK%20Youth)/Jones-%20The%20Youth%20Divide%20Diverging%20paths%20to%20adulthood.pdf (Accessed: February 18 2021).

Joona, T. (2018) *Everyday life in the Arctic*. In a blog: Polar prediction matters. Available at: https://blogs.helmholtz.de/polarpredictionmatters/ (Accessed: February 18 2021).

Kaiser, N. (2011) *Mental health problems among the Swedish reindeer-herding Sami population: In perspective of intersectionality, organisational culture and acculturation*. Umeå: Umeå University, Faculty of Medicine, Department of Clinical Sciences, Psychiatry.

Kaiser, N., Näckter, S., Karlsson, M. and Salander Renberg, E. (2015) 'Experiences of being a young female Sami reindeer herder: A qualitative study from the perspective of mental health and intersectionality', *Journal of Northern Studies, 9*(2), pp. 55–72. Available at: http://urn.kb.se/resolve?urn=urn:nbn:se:umu:diva-119918 (Accessed: February 18 2021).

Kaiser, N., Ruong, T. and Salander Renberg, E. (2013) 'Experiences of being a young male Sami reindeer herder: a qualitative study in perspective of mental health', *International Journal of Circumpolar Health, 72*(1). doi: 10.3402/ijch.v72i0.20926.

Kaiser, N., Sjölander Liljegren, A. E., Jacobsson, L. and Salander Renberg, E. (2010) 'Depression and anxiety in the reindeer-herding Sami population of Sweden', *International Journal of Circumpolar Health, 69*(4), pp. 383–393. doi: 10.3402/ijch.v69i4.17674.

Karlsdottir, A. (2015) 'Nordic youth and future visions', in Karlsdóttir, A. and Jungsberg, L. (eds.) *Nordic Arctic youth future perspectives*. Stockholm: Nordregio, pp. 65–70. Available at: http://norden.diva-portal.org/smash/get/diva2:1128959/FULLTEXT01.pdf (Accessed: February 19 2021).

Karlsdottir, A. and Jungsberg, L. (2015) Youth perspectives on their future in the Nordic Arctic. *Nordregio Policy Brief*, pp. 1–4. Available at: https://www.diva-portal.org/smash/get/diva2:843858/FULLTEXT01.pdf (Accessed: February 18 2021).

Keskitalo, P. (2019) 'Nomadic narratives of Sámi people's migration in historic and modern times', in Uusiautti, S. and Yeasmin, N. (eds.) *Human migration in the Arctic: the past, present, and future*. Singapore: Palgrave Macmillan, pp. 31–65.

Keskitalo, P., Frangou, S.-M., and Chohan, I. (2020) 'Educational design research in collaboration with students: developing a reindeer herding study programme and a model of vocational Sámi pedagogy', *Education in the North, 27*(1), pp. 58–77. doi: 10.26203/3jtv-9g81.

Kokkinen, A. (2012) *On Finland's economic growth and convergence with Sweden and the EU15 in the 20th century (Research Reports 258)*. Helsinki: Statistics Finland.

Kumpula, J. (2012) 'Ilmastonmuutos ja poronhoito', in Ruuhela, R. (ed.) *Miten väistämättömään ilmastonmuutokseen voidaan sopeutua? Yhteenveto suomalaisesta sopeutumistutkimuksesta eri toimialoilla'.* Maa- ja metsätalousministeriön julkaisuja, 6, 2011. Helsinki, Maa- ja metsätalousministeriö, pp. 56–60.

Laitala, M. and Puuronen, V. (2015) 'A day in my life as a 35-year-old', in Karlsdóttir, A. and Jungsberg, L. (eds.) *Nordic Arctic youth future perspectives.* Stockholm: Nordic Arctic Youth Future Perspectives, pp. 27–35. Available at: http://norden. diva-portal.org/smash/get/diva2:1128959/FULLTEXT01.pdf (Accessed: 19 February 2021).

Lépy, E., Heikkinen, H. I., Komu, T., and Sarkki, S. (2018) 'Participatory meaning making of environmental and cultural changes in reindeer herding in the northernmost border area of Sweden and Finland', *International Journal of Business and Globalisation, 20*(2). doi: 10.1504/IJBG.2018.089868.

Lindqvist, J. (2009) 'Reindeer herding: a traditional indigenous livelihood', *Macquarie Journal of International and Comparative Environmental Law, 6*(1), pp. 83–127.

Luke. (2016) *Porotalous* [Reindeer husbandry]. Available at: https://www.luke.fi/ tietoa-luonnonvaroista/maatalous-ja-maaseutu/porotalous/ (Accessed: February 19 2021).

Mbunda, Fr D. (1983) 'Cultural values, tradition and modernity', in UNESCO (ed.) *Problems of culture and cultural values in the contemporary world.* Paris: UNESCO, pp. 13–21. Available at:https://unesdoc.unesco.org/in/documentViewer.xhtml?v=2. 1.196&id=p::usmarcdef_0000054681&file=/in/rest/annotationSVC/ DownloadWatermarkedAttachment/attach_import_97ba803b-3251-4a9b-ae6e-860a8b2a888d%3F_%3D054681engo.pdf&locale=en&multi=true&ark=/ ark:/48223/pf0000054681/PDF/054681engo.pdf#%5B%7B%22num%22%3A43%2 C%22gen%22%3A0%7D%2C%7B%22name%22%3A%22XYZ%22%7D%2C-229%2C856%2C0%5D (Accessed: December 21 2020).

Merikivi, J., Myllyniemi, S., and Salasuo, M. (eds.) (2016) *Lasten ja nuorten vapaa-aikatutkimus 2016 mediasta ja liikunnasta. Media hanskassa* [Leisure survey of children and young people 2016 on media and exercise. Media in a glove]. Helsinki: Nuorisotutkimusseura. Available at: https://issuu.com/tietoanuorista/docs/lasten_ ja_nuorten_vapaa-aikatutkimu (Accessed: December 23 2021).

Miller, C. (2005) *Aspects of training that meet Indigenous Australians' aspirations: A systematic review of research.* Adelaide, Australia: National Centre for Vocational Education Research. Available at: https://files.eric.ed.gov/fulltext/ED493924.pdf (Accessed: November 25 2020).

Nieminen-Sundell, R. (2011) *Maisema on, työ puuttuu.* Helsinki: Sitra. Available at: https://media.sitra.fi/2017/02/27172821/Maisema_on_tyC3B6_puuttuu-2.pdf (Accessed: November 23 2020).

Nuorisolaki, 1285/2016. *Finlex.* Available at: https://www.finlex.fi/fi/laki/alkup/2016/ 20161285 (Accessed: July 15 2021).

Nymand Larsen, J., Fondahl, G., and Schweitzer, P. (2010). *Arctic social indicators: a follow-up to the Arctic Human Development Report.* Copenhagen: Nordic Council of Ministers. Available at: http://norden.diva-portal.org/smash/get/diva2:789051/ FULLTEXT02.pdf (Accessed: July 15 2021).

Öhman, M.-B. (2015) 'Sámi youth struggling for rights and recognition', in Karlsdóttir, A. and Jungsberg, L. (eds.) *Nordic Arctic youth future perspectives.* Stockholm: Nordic Arctic Youth Future Perspectives, pp. 47–54. Available at: http://norden. diva-portal.org/smash/get/diva2:1128959/FULLTEXT01.pdf (Accessed: October 29 2020).

Oinas, P. (2018) *Poroperheiden sosiaalinen ja taloudellinen selviytyminen elinkeinollisessa ja yhteisöllisessä murroksessa [Social and economic survival of reindeer families in the economic and communal transformation].* University of Lapland. Department of Social Sciences. Available at: https://lauda.ulapland.fi/bitstream/handle/10024/63156/Oinas.Pirjo.pdf?sequence=1&isAllowed=y (Accessed: October 29 2020).

Ojala, J., Eloranta, J. and Jalava, J. (eds.) (2006) *The road to prosperity: an economic history of Finland.* Helsinki: Suomalaisen Kirjallisuuden Seura.

Ollila, A. (2014) *Reindeer blog.* Rovaniemi: Paliskuntain yhdistys. Available at: https://paliskuntainyhdistys.blogspot.com/2014/08/ (Accessed: October 23 2020).

Ollila, A. (2019) *Poronhoitoa nyt ja tulevaisuudessa [Reindeer husbandry now and in the future].* Web log. Rovaniemi: Paliskuntain yhdistys. Available at: https://poromieslehti.blogspot.com/2019/12/poronhoitoa-nyt-ja-tulevaisuudessa.html (Accessed: October 23 2020).

Omma, L. M., Holmgren, L. E., and Jacobsson, L. H. (2011) 'Being a young Sami in Sweden: living conditions, identity and life satisfaction', *Journal of Northern Studies*, 5(1), pp. 9–28.

Pekkarinen, A. (2006) 'Changes in reindeer herding work and their effect on occupational accidents', *International Journal of Circumpolar Health*, 65(4), pp. 357–364. doi: 10.3402/ijch.v65i4.18125.

Pirttilä, I.-A. (2020) *18 years old reindeer herder girl Iida-Aletta and her Arctic life.* Visit Lapland. Available at: https://www.ourlapland.fi/reindeer-herder-girl-iida-aletta-from-lapland (Accessed: December 23 2020).

Pogodaev, M. and Oskal, A. (2015) *Youth. The future of reindeer herding peoples.* Documentation. Arctic Council. Available at: http://reindeerherding.org/wp-content/uploads/2015/01/ICRH-0115-01_Voice_of_Reindeer_Herding_Youth_v07.00.compressed.pdf.

Povoroznyuk, O., Habeck, J., and Vaté, V. (2010) 'Introduction: on the definition, theory and practice of gender shift in the north of Russia', *Anthropology of East Europe Review*, 28(2), pp. 1–37.

Rasmus, S., Kivinen, S., Bavay, M., & Heiskanen, J. (2016). Local and regional variability in snow conditions in northern Finland: a reindeer herding perspective, *Ambio*, 45(4), 398–414.

Rasmus, S., Turunen, M., Luomaranta, A., Kivinen, S., Jylhä, K., and Räihä, J. (2020) 'Climate change and reindeer management in Finland: Co-analysis of practitioner knowledge and meteorological data for better adaptation', *Science of the Total Environment*, 710. doi: 10.1016/j.scitotenv.2019.136229.

Rasmussen, R. O. (2009) 'Gender and generation: perspectives on ongoing social and environmental changes in the Arctic', *Signs*, 34(3), pp. 524–532. doi: https://doi.org/10.1086/593342.

Rehtonen, T. (2019) *Nuoria naisia ja vanhoja miehiä* [Young women and old men]. *Poromieslehti.* Availabe at: https://poromieslehti.blogspot.com/2019/11/nuoria-naisia-ja-vanhoja-miehia.html (Accessed: August 20 2020).

Reindeer Herders' Association. (2015) *Reindeer herding cooperatives.* Rovaniemi. Available at: https://paliskunnat.fi/reindeer/reindeer-herding/cooperatives/ (Accessed: 20 August 2020).

Reindeer Herders' Association. (2019) *Miesten ja naisten kehityksen määrä poronomistajissa* [The rate of development of men and women in reindeer herders]. Rovaniemi. Available at: https://paliskunnat.fi/py/materiaalit/tilastot/poronomistajat/miehia_ja_naisia_poronomistajissa_1990_2018/ (Accessed: August 20 2020).

Reindeer Herders' Association. (2020a) *Poro* [Reindeer]. Rovaniemi. Available at: https://paliskunnat.fi/py/wp-content/uploads/2020/09/Reindeer_2020.pdf (Accessed: August 18 2020).

Reindeer Herders' Association. (2020b) *Development of reindeer herding*. Rovaniemi. Available at: https://paliskunnat.fi/reindeer/reindeer-herding/history/ (Accessed: June 29 2020).

Reindeer Herders' Association. (2020c) *Poronuoret* [Reindeer youth]. Rovaniemi. Available at: https://paliskunnat.fi/poro/poronhoito/poromiehen-ammatti/poronuoret/ (Accessed: August 20 2020).

Reinert E. S., Aslaksen, I. Eira, I. M. G., Mathiesen, S. D., Reinert, H., and Turi, E. I. (2009) 'Adapting to climate change in Sámi reindeer herding: the nation-state as problem and solution', in Adger, W. N., Lorenzoni, I. and O'Brien, K. L. (eds.) *Adapting to climate change: Thresholds, values, governance*. Cambridge: Cambridge University Press, pp. 417–432.

Reiter, C. and Lutz, W. (2019) 'Survival and years of good life in Finland in the very long run', *Finnish Yearbook of Population Research*, *54*, pp. 1–27.

Sámi Education Institute. (2020) *Reindeer husbandry entrepreneurship: vocational qualification in nature and environment*. Inari. Available at: http://www.sogsakk. fi/en/Applicants/Educational-Training-Programs/Reindeer-Husbandry-Entrepreneurship (Accessed: August 20 2020).

Sarkki, S., Komu, T. Heikkinen, H. I., Acosta García, N., Lépy, É., and Herva, V.-P. (2016) 'Applying a synthetic approach to the resilience of Finnish reindeer herding as a changing livelihood', *Ecology and Society*, *21*(4). Available at: https://www. jstor.org/stable/26270038 (Accessed: November 20 2020).

Schindler, D. L. (1997) 'Redefining tradition and renegotiating ethnicity in native Russia', *Arctic Anthropology*, *34*(1), pp. 194–211.

Ssorin-Chaikov, N. V. (2003). *The social life of the state in subarctic Siberia*. Stanford: Stanford University Press.

Statistics Finland. (2019) *Finland among the best in the world*. Available at: http:// www.stat.fi/tup/satavuotias-suomi/suomi-maailmankarjessa_en.html (Accessed: August 20 2020).

Super, D. E., Savickas, M. L., and Super, C. M. (1996) 'The life-span approach to careers', *Career Choice and Development*, *3*, pp. 121–178.

Tveraa, T., Stien, A., Bårdsen, B.-J., and Fauchald, P. (2013) 'Population densities, vegetation green-up, and plant productivity: impacts on reproductive success and juvenile body mass in reindeer', *PLoS ONE*, *8*(2). doi: 10.1371/journal. pone.0056450.

Uboni, A. et al. (2016) 'Long-term trends and role of climate in the population dynamics of Eurasian reindeer', *PLoS ONE*, *11*(6). doi: 10.1371/journal. pone.0158359.

Ulturgasheva, O. (2012a) *Narrating the future in Siberia: childhood, adolescence and autobiography among Eveny*. Oxford: Berghahn Books. Available at: http://www. berghahnbooks.com/title.php?rowtag=UlturgashevaNarrating (Accessed: November 20 2020).

Ulturgasheva, O. (2012b) 'Navigating international, interdisciplinary, and Indigenous collaborative inquiry: Phase 1 in the circumpolar Indigenous pathways to adulthood project', *Journal Community Engagem Scholarsh*, *4*(1), pp. 50–59. Available at: https://digitalcommons.northgeorgia.edu/jces/vol4/iss1/6.

Ulturgasheva, O. (2014) 'Attaining khinem: Challenges, coping strategies and resilience among eveny adolescents in northeastern Siberia', *Transcultural Psychiatry*, *51*(5), pp. 632–650. doi: 10.1177/1363461514546246.

Ulturgasheva, O. (2015) 'Collapsing the distance: Indigenous-youth engagement in a circumpolar study of youth resilience', *Arctic Anthropology, 52*(1), pp. 60–70. doi: 10.3368/aa.52.1.60.

Ulturgasheva, O., Rasmus, S., Wexler, L., Nystad, K. and Kral, M. (2014) 'Arctic indigenous youth resilience and vulnerability: comparative analysis of adolescent experiences across five circumpolar communities', *Transcultural Psychiatry, 51*(5), pp. 735–756. doi: 10.1177/1363461514547120.

United Nations. (2009) *State of the world's Indigenous peoples*. New York, NY: Department of Economic and Social Affairs Division for Social Policy and Development Secretariat of the Permanent Forum on Indigenous Issues. Available at: https://www.un.org/esa/socdev/unpfii/documents/SOWIP/en/SOWIP_web.pdf (Accessed: October 29 2020).

Veijola, S. and Strauss-Mazzullo, H. (2018) 'Tourism at the crossroads of contesting paradigm of Arctic development', in Finger, M. and Heinonen, L. (eds.) *The Global Arctic Handbook*. Cham: Springer, pp. 63–81.

Vitebsky, P. (2010) 'From materfamilias to dinner-lady: The administrative destruction of the reindeer herder's family life', *The Anthropology of East Europe Review, 28*, pp. 38–50.

Voutilainen, M. (2016) *Poverty, inequality and the Finnish 1860s famine*. Doctoral dissertation, University of Jyväskylä. Available at: https://jyx.jyu.fi/bitstream/handle/123456789/49598/978-951-39-6627-0_vaitos13052016.pdf?sequence=1.

Vuojala-Magga, T. and Turunen, M. T. (2015) 'Sámi reindeer herders' perspective on herbivory of subarctic mountain birch forests by geometrid moths and reindeer: a case study from northernmost Finland', *Springerplus, 4*(134). Available at: https://www.ncbi.nlm.nih.gov/pmc/articles/PMC4374085/ (Accessed: November 14 2020).

Vuojala-Magga, T. et al. (2011) 'Resonance strategies of Sámi reindeer herders in northernmost Finland during climatically extreme years', *Arctic, 64*(2), pp. 227–241. Available at: www.jstor.org/stable/23025696 (Accessed: August 20 2020).

Wennstedt, E. B. (2002) 'Reindeer herding and history in the mountains of Southern Sapmi', *Current Swedish Archaeology, 10*, pp. 115–136. Available at:http://www.arkeologiskasamfundet.se/csa/Dokument/Volumes/csa_vol_10_2002/csa_vol_10_2002_s115-136_wennstedt-edvinger.pdf (Accessed: November 20 2020).

West, P., Sweeting, H., and Young, R. (2010) 'Transition matters: pupils' experiences of the primary-secondary school transition in the West of Scotland and consequences for well-being', *Research Papers in Education, 25*(1), pp. 21–50. doi: 10.1080/02671520802308677.

Appendix

Interview Questions
Age:
Gender:
Education:
What is your first memory related to reindeer herding?
Is reindeer herding practised in your family? Who is the most active member?

Do you know how long reindeer herding has been practised in your family?
How do you participate in reindeer herding and related work?
What is the most interesting (the most fun) thing when you're involved in reindeer herding?
What are your future plans (after completing school)?
Do you want to study more?
Would you like to continue in reindeer herding? How? Would it be possible to combine different occupations?
Do you want to continue living where you live right now, or do you want to explore the world a little bit? Study/work somewhere else?
What do you think about the future of reindeer herding?
What do you think about climate change? Do you think it will affect reindeer herding?
What do you think about reindeer herding in general? Do you think that reindeer herding will bring you income in the future?

Interview questions for guardians:
What do you think of the future of reindeer herding related to your children?
Can you describe the different kinds of tasks that the family does?

6   Indigenous youth perspectives on extractivism and living in a good way in the Yukon

Susanna Gartler with Taiya Melancon and Eileen Peter

Setting the stage

The Yukon (see Figure 6.1) is a part of Canada's Arctic boreal forest, often portrayed as *"a biological treasure, Indigenous homeland, and extractivist frontier"* (Willow 2016, p. 1). This 'frontier' exists because of the Tintina gold belt, stretching from Alaska to the Yukon: a geological region rich in not only gold and silver, but also copper, lead, zinc, tungsten, and uranium (Goldfarb et al., 2000). Mineral production quickly became industrialized after the gold and silver rushes of the late nineteenth and early twentieth century (Coates & Morrison 2005). During this history of gold and galena extraction, settlers and the Indigenous population became *"irrevocably intertwined"* (Winton & Hogan 2015, p.93). The present case study asks how problems Indigenous youth in the Yukon are confronted with are intricately tied to extractivism (Klein 2011; Acosta 2013; Petras and Veltmeyer 2014)—a defining characteristic of the Canadian settler state (Bélanger 2018). Extractivism, as we employ it here, is the continuation of environmentally and socially disruptive coloniality (Willow 2016). By focusing on a resilience approach (Roe et al. 2012), this chapter sketches ideas and solutions proposed by the study participants for 'living in a good way.' Living in a good way in a Yukon First Nation context means following ethical and cultural protocols and contributing to the wellbeing of oneself and one's community (Demientieff 2017).

Indigenous peoples of the Yukon have lived on their traditional territories since 'time immemorial.' After contact with fur traders and prospectors in the late nineteenth century, Yukon First Nations gradually became part of the cash economy, while continuing to live off the land (Coates and Morrison 2005). Today, through land claims and self-government agreements, Yukon First Nations retain ownership to parts of their traditional homelands.[1] Referring to the large and still unremediated mine sites near Mayo, called Elsa and Keno Hills, Winton and Hogan (2015) find that

> Knit together in a complex pattern of mutual involvement and unequal impacts, the story of the Na-Cho Nyäk Dun and the Keno Hill mine is illustrative of how Aboriginal people across Northern Canada have been

DOI: 10.4324/9781003110019-9

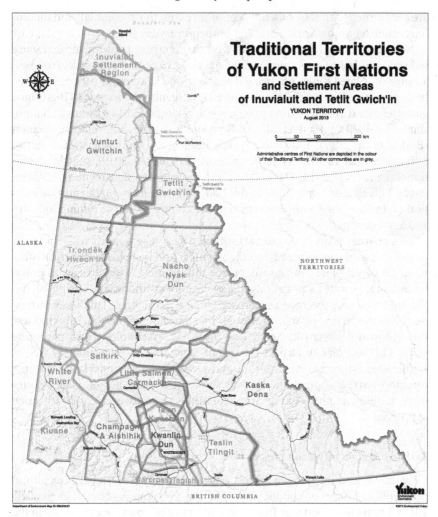

Figure 6.1 Traditional Territories of Yukon First Nations, and Settlement Areas of Inuvialuit & Tetlit Gwich'in.

(Source: Government of Yukon)

both affected by and involved in one of the most destructive forms of industrial development in their traditional lands.

(Winton and Hogan, 2015, p. 93)

The repercussions of this accumulation by dispossession (Harvey 2003) can still be felt today, including the slow violence (Nixon 2011; Sandlos and Keeling 2016) of environmental contamination. While mining operations today benefit the local population primarily by providing job opportunities, this chapter shows that First Nation youth, and young women in particular, are confronted with the consequences of an extractivist, colonial, and masculinist mindset at the work place and in their everyday lives.

Not many studies focus specifically on the interplay between the extractive industry and Indigenous youth wellbeing in the Arctic and Subarctic. However, a range of related topics are being discussed, which, amongst others, form the basis of Indigenous wellbeing, such as environmental repossession (Big-Canoe and Richmond 2014), sharing practices (Rasmus et al. 2014), positive relationships (Kral 2012; Kral et al. 2014), community-led cultural revitalization (Barker et al. 2017), Indigenous youth resilience (Ulturgasheva et al. 2014; Ulturgasheva et al. 2015), safer-sex efficacy (Logie et al. 2019), suicide prevention (Rasmus et al. 2014; Talaga 2020), and Indigenous - state relations (Dhillon 2017). This chapter contributes to this body of literature by highlighting a selection of Indigenous youth concerns that are connected to the mining industry and a colonial past.

This chapter is divided into two sub-sections: 'There needs to be balance' describes problems connected to the labour market, such as worker satisfaction, gendered biases in the workplace and (coping with) drug abuse and addictions, as well as protecting the environment and revitalization of on-the-land skills in the face of climate change. 'Northern Tutchone Strong' discusses fostering intergenerational understanding and lateral kindness through youth initiatives as well as the use of technology and parenting skills. This section emphasizes the role of education and schooling as well as Indigenous languages and culture and strengthening connections through ceremony, art, and play. The chapter shows how these, at first glance seemingly unrelated, phenomena are structurally tied to each other through extractivism.

Living in a good way and extractivism

Generally speaking, living in a good way means following ethical and cultural protocols and codes of conduct in Yukon First Nations contexts (Demientieff 2017). Living in a good way from the perspective of young people living in the Yukon also means doing good things, in the right way. A 27-year-old entrepreneur and youth advocate—who self-identifies as Ojibwe, Northern Tutchone and Kaska Dene—explains: "You always hear that in Indigenous culture everywhere, that we need to live in a good way. You do good things; good things will come." A concrete example would also be the Ojibwe/ Anishinaabe concept of Pimaa-tisiwin, which *"does not simply imply an accumulation of 'good' things for the self, but also an ethic of goodness"* (Kulchyski 2005, p. 113). Living in a good way and Indigenous wellbeing is intricately tied to connection to ancestral land (Bhattacharyya et al. 2013; Petrasek MacDonald et al. 2015).

Extractivism—also coined "extractive imperialism" by Petras and Veltmeyer (2014)—disrupts Indigenous connection to land, and is based on the export of primary commodities (Acosta 2013), associated with high poverty levels, uneven wealth distribution and imperialist systems (Willow 2016).

Defined "*as an environmentally and socially destructive extension of an enduring colonial societal structure*" (Willow 2016, p. 1), Klein (2011) notes how
extractivism is bound up with the idea of a limitless nature. Sehlin MacNeil
(2017) develops the concept of extractive violence in relation to Indigenous
peoples' "*as violence against people and/or animals and nature caused by
extractivism, which predominantly impacts peoples closely connected to land*"
(ibid., 2). Extractivism "*reproduces the resource colonialism of old*" (Willow
2016:13), benefiting those most who are already empowered—while locals
bear the most of the environmental and social risks (Burke 2012).

Methods and methodology

Following Indigenous methodologies of reflexivity and accountability (Chilisa
2012; Russell-Mundine 2012; Smith 2012; Rasmus 2014), I will briefly present
my own background: Born in Austria to an Austrian father and a Czech mother,
I studied Social and Cultural Anthropology at the University of Vienna, where
I am currently enrolled in a doctoral programme. My previous engagement
with subsistence, land use and Indigenous self-determination (Gartler 2018)
engendered my interest in the Arctic and Subarctic. Researchers working in the
Yukon are required to obtain a research license and ethics approval prior to
commencing their work and First Nations' permission to conduct fieldwork.
Collaboration with First Nations is strongly encouraged, enabling co-design of
research aims and methodologies. Between 2014 and 2019, I was co-investigator
of the community-based participatory project[2] "LACE—Labour Mobility and
Community Participation in the Extractive Industry Case Study in the Yukon
Territory (2014–2019)" (Saxinger 2018) and worked with the Heritage
Department of the First Nation of Nacho Nyäk Dun (FN NND) on a variety
of themes, including oral history, language revitalization, preservation of heritage sites, and the planning of a cultural centre.[3]

This chapter is based on participant observation at a variety of Indigenous
youth and cultural events and informal conversations conducted throughout
2014 to 2019. It is also based on semi-structured anonymous interviews conducted within the framework of LACE with six Indigenous mine workers.
Moreover, two First Nation of Nacho Nyäk Dun youth councillors[4], one
young member of the Heritage Department and a FN NND community education liaison coordinator (now former positions), as well as an Indigenous
youth advocate participated. Data collection also included co-creation of
knowledge through research-topic and collaborative yarning (Bessarab and
Ng'Andu 2010) with two young FN NND citizens (one male, one female),
and one semi-structured interview with a non-Indigenous female, as part of
the project Arctic Youth and Sustainable Futures.[5] All participants were
asked to review an earlier version of this chapter and whether they wanted to
be co-authors. Two of them, Taiya Melancon and Eileen Peter (both FN
NND citizens), accepted the invitation.

At the time of the interviews all participants were under the age of 30. The
research sampling of participants between the ages of 18 and 30 corresponds

to local definitions of youth (see e.g., Council of Yukon First Nations 2020). However, First Nation definitions are more complex, as this participant explains:

> As I have always been taught, you are a young person until you find out what your gifts are. When we find those gifts, our job is to give things away—and then we find our journey. And when we find our journey, we become an adult. These are the steps in life that we take.

These thoughts resonate with Indigenous scholar Marcus (2016, p. 359) who explains that being aware of your gifts and talents entails a sense of being in place, socially, and physically: "*By being mindful of the gifts of our lives within our own talents and abilities, we become more attuned to our place within our community and the environment.*" Most interviews took place on the traditional territory of the First Nation of Nacho Nyäk Dun, and on the traditional territory of the Kwanlin Dün First Nation and the Ta'an Kwäch'än Council. We wholeheartedly subscribe to the call to challenge the "*deficit framing of the Aboriginal situation*" (Parlee and Marlowe 2002): This chapter thus addresses not only problems Indigenous youth and young adults in the Yukon face but also discusses how to resolve these problems—thus employing a resilience and asset approach (Roe et al. 2012).

"There needs to be balance": protecting nature and climate, jobs and the work environment

The two main drivers of the Arctic economy today are subsistence harvesting and the extraction of renewable and non-renewable natural resources (Holen et al. 2015; Larsen and Huskey 2020), which are mostly sold to regions far South of the subarctic boreal forests. Resource revenues from mineral extraction are difficult to estimate, but research has shown that local financial benefits have been limited in the past (Huskey and Southcott 2016). Prices are generally higher in remote northern communities, and housing can become scarce due to the industry's demands. However, mining brings benefits too, such as employment, scholarships and royalties (especially if the mine is operating on Category A land).

General opinions around mining usually revolve around two issues in the Yukon: Jobs and the Environment. This young miner echoes the opinion of many, including First Nation of Nacho Nyäk Dun Elders (FN NND Elders et al. 2019): "Mining isn't a good or a bad thing really. Because you are destroying lots of stuff and you are polluting the water. At the same time, it has given people jobs and employment." In contrast, this youth advocate shares his relational view:

> My teachings, they say that anything that disturbs the land, unless you give back, isn't good. Everything underneath the earth is connected. So, when you disconnect what is underneath, you mess up the eco system and the relationship between you and the land.

Yukon Indigenous peoples traditionally consider themselves part of the land and part of the water (McClellan et al. 1987). He adds that "the least you can do is reclaiming the land: Make sure you return the land to how it was when you got there—considering you are going to do it either way." Balance is a concept that often comes up in connection to a reciprocal relationship with land (see e.g., Trinidad 2012) and mining, as this youth councillor says:

> My personal opinion is that it needs to be balanced. It brings a lot of opportunities to youth, and a lot of economic development, it generates income for families. But the mines also need to respect our opinion, work with us, talk with us and negotiate based on mutual respect.

Attitudes towards working in the industry differ from person to person: some perceive the working conditions of rotational shift work as overpowering; others adjust to it more easily (Saxinger and Gartler 2017). Experienced miners recount that on-the-job, step-by-step learning and mentoring—as well as being able to adjust to the boom and bust cycles of mining—is essential (Saxinger and Gartler 2017). However, Indigenous women and men are often employed in low-paying, entry-level jobs; higher-ranking positions are often filled by experienced workers from outside the Yukon. Even fewer young Indigenous female—or 2SLGBTQQIA—persons are currently in leadership positions in the industry. Some Indigenous young people enjoy shift-work, are able to take advantage of the opportunity to work for high wages, and build their professional careers over time, whether through direct employment or by building their own companies. Two young members of the FN NND offer a differential view on the effects of the mining industry on the job market. When asked about work opportunities—at a time when a large gold mine had already started operating in the vicinity of her hometown—one answers promptly: "There's tons of jobs. If you want to work, you can." Her friend, however, remarks: "Yeah, but the jobs are only in the mining industry. It's where you get the most jobs with the highest pay." Reflecting on the very specific social and spatial setting of mine camps with strict rules and regulations (see Saxinger et al. 2016) he adds: "It's like going to prison." Other youth, however, do not see routines at camp and personal freedom as contradictory, suggesting that satisfaction with this type of work is dependent on individual preference—as well as family background. One participant, whose male family members have worked in mining for generations, responds by affirming that good camp food and good wages contribute significantly to worker satisfaction: "Yeah, but you get paid a lot. And the food is really good!"

Having access to locally hunted and harvested foods is an important part of living in a good way in the Yukon. However, large-scale mining operations often not only go hand in hand with accumulation by dispossession (Holden et al. 2011) but also with *the suppression of alternative (Indigenous) forms of production and consumption*" (Harvey 2006, p. 153). Good wages often allow workers to buy hunting, fishing and trapping equipment and long periods at home often allow for enough time for these activities

(Saxinger and Gartler 2017). At the same time, rotational shift work can present significant barriers to: (1) being on the land, by taking away time for hunting trips (Nelson et al. 2005); (2) participating meaningfully in community life; as well as (3) meeting childcare obligations (NIIMMIWG 2020, p. 593). A young worker explains: "When I come home from my shift I want to be with my family, go trapping and hunting, do my sewing and beading, all those things I don't have time for when I'm away."

Combining rotational shift work with being on the land, community, and family life is essential for living in a good way for Indigenous (and many non-Indigenous) workers. Through warming weather and other local impacts of climate change, however, it becomes harder to navigate safely on the land and to correctly assess indicators of risky conditions (Laidler et al. 2009, p. 388). In addition, several study participants noted that as on-the-land skills are being pushed to the margin by today's consumer society, skills such as navigating without instruments, dressing appropriately, and knowledge of the right equipment are increasingly lost. Thus, local impacts of climate change and the legacy of an extractivist system, which systematically devalued Indigenous ways of knowing, makes moving on the land and water increasingly difficult and unpredictable (see e.g., IRC 2016). Capitalogenic global warming, caused by excessive extraction, misappropriation and irresponsible use of resources (Harvey 2007; Malm and Hornborg 2014; Moore 2015; Bartolovich 2019) is also raised by a young man, who recently participated in Tracking Change, a program that brought him to the UN Climate Conference COP24 in Poland. He asserts what has become common knowledge in the past years: "Climate change is happening in the Arctic and Subarctic, and it's three times faster here than in most places" (see also Bush and Lemmen 2019).

Learning how to adapt to a changing environment by spending more time on the land is a major concern for youth living in the Canadian Arctic. The Indigenous Youth Podium Discussion at the Adäka Festival in Whitehorse in 2016 addressed how to incorporate more Indigenous ways of being into their lives while being integrated in the modern world. One participant explains: "Our elders are saying now that what we have done to the world, we cannot take back. We now have to live with what we have done, but we have to be positive as young people." In response to these issues, in late 2020, the Council of Yukon First Nations, the Assembly of First Nations and the Yukon Region and Youth Climate Lab launched a call for 12 Youth Yukon First Nation Climate Action Fellowships.

Addictions, sexism, and workplace harassment

Extractivism has not only had profoundly negative impacts on the environment, mining colonialism has also gone hand in hand with assimilationist policies, the repercussions of which pose significant barriers for Indigenous youth to become resilient and strong leaders. Substance abuse, bullying, lateral violence and intergenerational blame which are linked to the Canadian

Residential School system (Bombay et al. 2014), can be regarded as part of an extractivist past and present too. Two young adults agree that their generation wants to break out of the cycle: "We see it all around us, but we don't want to be like that." Some youth are speaking on social media about how abuse of drugs is affecting their families and communities. One participant explains why addiction matters to her—a problem commonly associated with communities near mines (see, e.g., Gibson and Klinck 2005): "Drugs and the problems associated with that are real, maybe that is why I talk like that, why I am so angry." However, for some, selling drugs is a way to ensure an income in an environment where not many opportunities present themselves, as well as during recessions or mining busts. A young employee at a mine admits:

> A few years ago, during the world economic crises, I got involved with a gang and sold a lot of drugs. They are worth way more in the North, so I moved here. But then I got a job really quickly, and slowly fell away from that. So, work has been really positive in my life, helping me with my problems.

Other young mine employees share that being in camp for two weeks helps them get away from their everyday problems and (bad) habits. Another participant, who himself went through alcohol abuse and was able to heal himself, shares that ways of helping peers are numerous, such as creating your own organization, or simply talking to people, making sure they have someone to confide in.

Trust and confidence are important elements of workplace safety. In Canada, female employment in mining makes up 14% of the workforce (Natural Resources Canada 2019). Not all experiences of young women—whether Indigenous or non-Indigenous—are good though, and sexist behaviours and assumptions still prevail in an *"often hypermasculine and hypersexualized"* (NIIMMIWG 2020, p. 593) work environment, at both the blue- and white-collar levels. A 27-year-old non-Indigenous Canadian professional working closely with industry proponents, explains:

> Drillers and other trades people do a lot of 'boys talk'. It's very direct, as a woman you have to adapt to the male environment. The business man coming from the South will be subtler, carrying sexist assumptions, thinking you are there to take notes, or ignoring you. It's like a 'boys can talk to the boys' club. Men in the (mining) industry still have a hard time acknowledging that—especially young—women can be in leadership roles.

A young Indigenous housekeeper at a mine camp notes that harassment can be rather subtle too: "I have an admirer out here he just likes to stare at me. It's really annoying. If he doesn't stop soon, I'll have to say something." However, harassment is not confined to camps and happens in communities too. The connection between violence, colonialism, land rights, mining, female bodies, and youth was observed a very long time ago by Indigenous activists, as Clark (2016, p. 49) notes:

> [...] early Indigenous activists such as Zitkala-Sa and Winnemucca (1883) were central to fighting the issues of violence on the land and on the body as they witnessed it at the turn of the century. [...] Zitkala-Sa and other Indigenous feminists remind us again and again in their writing that violence has always been gendered, aged, and linked to access to land.

The connections between resource extraction, substance abuse, and violence against Indigenous women, girls and 2SLGBTQQIA[6] people has been pointed out in the Report of the National Inquiry Into Missing and Murdered Indigenous Women and Girls (NIIMMIWG 2020). Linking extractivism and the (failure of) recognition of Indigenous rights to violence against 2SLGBTQQIA people, Indigenous women, and girls, the report calls upon the extractive industry and its workers to address *"situations in which violence is [perpetrated,] ignored and normalized as part of the work environment"* (NIIMMIWG 2020, p. 593)—among many other recommendations. Companies in the Yukon are required to have strict policies to deal with perpetrators, who are—ideally but not always—terminated and sent home immediately if an incident occurs. Other coping mechanisms include enlisting allies, support circles, using media to speak up, and informing the public or management of the company if any transgressions occur.

A youth advocate stresses the need for strong future leaders to implement necessary changes for environmental protection and a safer work and home environment for all: "We need to make the younger people stronger, so that we will have future leaders that are going to uphold the protection of our people, the land, water and the air." Big-Canoe and Richmond (2014) call for 'environmental repossession' as a response to disconnection from and dispossession of land. In response to the court case surrounding the Peel Water Shed, the FN NND has organized several "Youth of the Peel"[7] canoe trips in recent years, to promote strong ties to land, culture and community (Petrasek MacDonald et al. 2015), and to promote knowledge transfer between Elders and youth. They also sent some youth to the Supreme Court in Ottawa for the final decision in the case. In the words of a youth councillor, this was

> to show young people why we do what we do and why are we protecting this area from mining and exploration. You got to show them one on one, also how far we will go to protect something we believe in.

She explains,

> Being Northern Tutchone in my own view means knowing the land, respecting it and treating everyone with respect. We say we are the

protectors of the land and the water. That's what being Northern Tutchone means to me.

Positions such as the youth councillor within a First Nation government are designed to build confidence and leadership skills. On-the-land initiatives such as Youth of the Peel and cultural camps include the transmission of traditional laws—such as respecting Elders and living in a good way, according to Doo'Lí laws (Natcher and Davis 2007) and the Northern Tutchone Code of Ethics—and knowing how to hunt, trap, fish and gather, and process plants and medicine. At the same time, young Indigenous adults take matters into their own hands as this story, shared by the young heritage and culture worker, shows:

> We formed a group, it's called Yukon First Nation Emerging Leaders[8] and that was made by young people for young people. And we organized the first ever Yukon First Nation Youth Conference that brought together youth from the whole Yukon. It was a three-day camping trip; we made sure it was on the land and it was amazing.

A variety of youth-initiated and First Nation programs and activities foster healthy living, intergenerational understanding and lateral kindness, including 'Youth for Lateral Kindness' founded by two members of the Kwanlin Dun First Nation in 2016 (CBC News 2017). Many other initiatives foster leadership, such as First Nation Youth Gatherings, the Yukon Youth Healthcare Summit (which won the Arctic Inspiration Price in February 2020), Yukon Youth organizations such as BYTE (see yukonyouth.com), sports events like the Yukon Native Hockey Tournament, education and career support projects like Northern Compass, and the Arctic Indigenous Youth Leaders' Summit, taking place in 2019. The next section deals with some barriers to effective leadership, such as grief, lateral violence, and intergenerational blame, and how they affect young Indigenous people's lives—while suggesting ways to overcome these structurally embedded effects of extractivism.

"Northern Tutchone Strong": language revitalization, education, ceremony, art and play

> Knowledge we regain, culture is my life
> Smudgin' on the sage, that's my right
> Are we on the same page, culture take flight
> Sitting in class, no end in sight
> Just want to be outside, out where them trouts dive
> Me and my homie Jeremiah gonna fly
> We can tell our kids we didn't let our culture die
> Strive to survive, keep our language alive
> (Song text by Driven to Change 2018)[9]

Community- and youth-led cultural revitalization is generally acknowledged as crucial to enhance Indigenous youth resilience and wellbeing (e.g., Big-Canoe and Richmond 2014; Ulturgasheva et al. 2014; Hatala et al. 2017), and also in relation to a pandemic of another kind: youth suicide (Barker et al. 2017). Suicide disproportionately affects Indigenous boys and young men but also girls and young women in Canada (Kral 2012, 2013). An abrupt death and a lack of coping mechanisms within the family and social group can put enormous pressure on young people. A youth councillor summarizes preventive measures:

> We are putting a lot of effort into wellness, but sometimes there are unfortunate events that trigger a downward spiral. We are such a small community, sometimes there is nothing to do and that's where it all starts. People need opportunities to work, to have access to housing and to take part in land-based initiatives: these are the three things to keep our people healthy. And a good, low-threshold, easy-to-access support system, with councillors who live in town.

In the case of Mayo, these challenges are taken on by the community as a whole, but they require a very sensitive and open engagement with the colonial past and present, which is unfortunately not always the case. Opinions amongst some non-indigenous Yukoners, such as that First Nation's people should 'finally get over it', still exist. However, this kind of opinion does not acknowledge the communal effort it takes to reconcile with a violent past: If the inflicted violence is rooted in collectively held opinions and acts, only combinations of individual and collective approaches to healing can afford results (see, e.g., Lavallee and Poole 2010).

Language revitalization is an integral part of Indigenous resurgence. Language, as a means of expressing cultural knowledge (Marcus 2016), is "one of the biggest things that we are trying to revitalize everywhere", in the words of a youth advocate. Northern Tutchone is being taught at the local school in Mayo, but participants in the study stress the room for improvement. First Nations languages are key to re-indigenizing life in the North and should be taught at all levels starting at the day-care, a young FN NND citizen declares. His friend adds that "it's harder for us to revitalize our language and culture, when we don't really know how. But we know that there are things like language nests and immersion programmes, which work really well in other places." The two reflect on how revitalizing language and culture is much harder in a place so heavily affected by mining, however: "This town is definitely more colonized than any other place, because this place was built around a mining community." Other communities which don't have mines "definitely have more First Nations who know what they're talking about and who know what they're doing. So, it's kind of difficult for us. Because we're so … adapted." The short pause she adds before the word 'adapted' points to a salient issue: Adaptation and flexibility are attributes commonly associated with First Nation, Inuit and Métis peoples (see e.g., Gartler et al. 2019).

However, if Indigenous peoples across Canada would have been able to adapt on their own terms to settler society—instead of being forcibly assimilated—the 'Settler problem' (Taiaiake Alfred in Regan 2010, p. x) would have been much less aggravated than it is today.

Mining provides the resources to build 'brand-new trucks' (a very popular commodity in the Yukon, also among Youth), ATV's, skidoos, other outdoor equipment as well as the electronic gadgets the younger generations are so fond of and accustomed to. While younger children play outside, it is often hard to pry teenagers and young adults away from their electronic devices. The excessive use of technology is frequently called out as a problem by youth and young adults themselves, although there are not just negative aspects to technology: snowmobiles and GPS, for example, enable young hunters to pursue subsistence activities without the years of experience needed to deploy a dog team or navigate using traditional methods (Laidler et al. 2009, p. 389). Moreover, youth agree that taking advantage of information technology to bring back cultural understandings and learning First Nation languages is very useful.

Whether through individual or collective action, face-to-face interaction or using the technological possibilities now available, Indigenous youth identify revitalizing culture and language as an integral part of the path towards healing and achieving community wellbeing. Part of Indigenous resurgence is an active stance against the negative effects of the Canadian Residential School system and the acknowledgment that it did not all happen in the distant past. A youth councillor explains:

> Fighting the effects of residential schools is a slow process. Lots of people think it was a long time ago, but it's not, it's our parents' generation too. And we are just still fighting all the things that came from it.

Another major problem associated with the Canadian Residential School system, which is identified by both a FN NND youth councillor and a long-term youth advocate, is parenting skills. In response to this problem, a Parenting Support Group was founded by a youth advocate and two other individuals in Mayo to give support to young parents and single mothers.

A youth councillor explains:

> First Nation people are slowly building themselves back and getting back to how we were before colonialism. This includes having things that grounded us, like to get our language and culture back, that will create a better future for our people. And we try to build up healthy young aboriginal people, what we call 'Northern Tutchone Strong'—having strong individuals, that can fight for our people in the future. And not only our people, but First Nations in general.

When asked how the younger generation envision their future, the member of the FN NND Heritage and Culture department explains:

> Less talking—more action. In the office here you get caught up with paper work so much and talking about things, instead of just doing it. I think youth are more action based. They brought so much energy and ideas to the Yukon First Nation Youth Conference I was part of this year.

Her statement mirrors the grievance that Yukon First Nations had when building government structures in accordance with the colonial system in order to be legitimate for the Canadian state and enter treaty negotiations, thereby ironically undermining long-standing principles of Indigenous governance (Nadasdy 2017). Moreover, as gold and other mineral prices skyrocket during economic crises (such as after 2008 and again in 2020), First Nation self-governments can become inundated with requests from mining proponents to claim stakes and conduct exploration on their traditional territories. A youth advocates expresses this as follows: "We are using the system now that was designed to destroy us. That is so misguided." Governance structures thus need to be revitalized alongside language and all other aspects of Yukon First Nation cultures, such as education.

Indigenous education and schooling have a long and difficult history in Canada, and exemplify the destructive effects of a colonial, extractivist history and its effects on Indigenous identity. Recent detection of mass burials near former residential schools sparked renewed debate around Canada's attempted genocide of First Nations, Métis and Inuit peoples and brought the issue to the attention of international audiences (Assembly of First Nations 2021). Schooling today has been identified both as a resource for youth wellbeing as well as a source of challenges in the Arctic (Ulturgasheva et al. 2014, p. 739). "I think the school system is challenging" observes one youth, and adds that "teachers should be trained on the history of aboriginal people and what comes along with that". Her friend concurs and remarks upon missing qualifications of teachers as well as the lack of Indigenous teachers. The young woman explains that she wants her grades to reflect her actual skills, and that she feels she would be getting a better grade "learning about my history and First Nations. That's where it's easy to connect to for me." Learning about place and their traditional territories is part of the processes of reinhabitation and indigenization within the Critical Indigenous Pedagogy of Place approach (Trinidad 2012). An important aspect of this approach, and the traditional style of education among Yukon First Nations, is to learn by listening, watching, and doing on the land. Although First Nation content is actively integrated into the school curriculum through a FN NND Community Education Liaison Coordinator in Mayo, he notes that how much happens depends on teachers and how open they are to the First Nation curriculum:

> I like to bring in Elders and organize a lot of events here at the school. Recently, we had drum making, we have trapping, hunting, and fishing programs, and culture camps. But I believe that it's never enough, while at the same time a lot of these classes are trying really hard.

Marcus (2016, p. 356) explains that *"[f]or traditional Indigenous communities, ceremony is a way of life."* Indigenous ceremonial and spiritual practices such as smudging or offering tobacco, and arts like painting, weaving, storytelling, carving, beading and sewing, singing, dancing, and playing instruments, alone or in groups such as the women's drum group in Mayo called 'Rammi Elin', the Dakwäkäda Dancers or the Selkirk Spirit Dancers—including ceremonial components embedded in these practices—are increasingly popular and inspire youth from the early ages onward. After decades of repression, the tide might be turning again; the young member of the FN NND Heritage and Culture Department explains:

> The new generation, the young kids, they are so hungry and interested in those types of things. At culture camps, all the kids are trying to bead and cut the fish and tan the hides. They are right in there.

Young adults also actively reclaim the way their ancestors lived: "I still go hunting. Last time, to honor my grandma, I did it the traditional way, by walking." The First Nation of Nacho Nyäk Dun encourages youth and young adults to participate in activities and many actively seek to learn traditional practices, in their homes, during culture weeks and camps, and as part of the First Nation school curriculum. Other traditions are being actively promoted too, such as the 'First Hunt', which initiates a young man (or woman nowadays) into adulthood, according to a FN NND Community Education Liaison Coordinator. Rasmus et al. (2014) emphasize the importance of support groups and sharing practices over market relations for youth wellbeing in Arctic communities (see also Ulturgasheva et al. 2015, p. 752). If the First Hunt is successful, the caribou or moose meat is then shared with the wider community and especially Elders.

Being on the land and cultural practices foster what Marcus (2016, p. 359) calls the guiding principles of relationship, respect, responsibility, and reciprocity, which are *"understood through Indigenous ceremony"*. Many young artists combine their artistic practice with a critique of the status quo and a focus on indigenizing, self-determination, exploring Indigenous forms of self-governance, as well as re-connecting to spiritual and ceremonial practices and beliefs (see also the quote at the beginning of this section of a rap song by Driven to Change 2018; and interviews with young artists in Gignac 2019; Liu 2019). On a scale from cultural embeddedness to disconnection 'the Yukon is coming back', a youth advocate rejoices. When young Indigenous adults speak of cultural embeddedness and connection the above-mentioned principles of relationship, respect, responsibility, and reciprocity come to mind. During traditional hand games, for example, youth and young adults come together to have fun and to celebrate such collectively held values (see Figure 6.2). One of the organizers explains: "It's more than just about winning the tournament, it's about the community and the environment. That's what our people represent as a community: we come and work together and help each other." A youth advocate confirms the overall spirit that can be felt during everyday life, and at

Figure 6.2 Hand games held at the Kwanlin Dun Cultural Center in Whitehorse in 2016 during the Adäka Festival.

(Photo: Susanna Gartler)

cultural events in the Yukon: "We are re-claiming who we were. We are trying to figure out who we are, where we are coming from—and we're revitalizing a lot of things. We are in a really beautiful state right now."

Discussion and conclusion

An extensive study on diamond mine development with Tłįchǫ youth in Behchokǫ̀, Northwest Territories, highlights various—not only negative—impacts of the extractive industry, such as increased population transience, tensions between work and education, mines as community resource, including a positive effect on future outlook, inequities, and particular views of schooling (Davison and Hawe 2012, p. 218). The authors explain that *"Contradictory findings from previous studies suggest that the impact of mining is neither easy to predict nor simple to reveal"* (Davison and Hawe 2012, p. 223). From existing literature, they note several negative impacts of mining on youth in Canada such as changes in social ties and health, more sexually transmitted diseases, impacts on language and culture, and interference with traditional food sources. Increased income and economic prosperity are often cited as positive effects, but can lead to an increase in alcohol and drug consumption as well as gambling. Moreover, while educational attainment may decrease during a boom, addictions can increase and housing often becomes a problem—all of which are concerns in part confirmed by the present study.

The present case study shows that in the Yukon, mining can provide an opportunity to grow careers for young adults, to make good money and to get away from problems associated with living in small towns—such as feelings of intense social proximity, closeness to problems of other community members or isolation, and a lack of access to land and sense of belonging. Moreover, high wages allow workers to buy equipment and can thus facilitate subsistence living (Saxinger and Gartler 2017). However, racism, sexism, violence and age-ism as well as intersectional challenges prevail, in an environment which often perpetuates toxic masculinities (see also NIIMMIWG 2020). Young women in higher positions might not be perceived as being in leadership roles, and young Indigenous women and men will often still be employed in entry-level, low-paying jobs. From the perspective of reciprocity, mining will always remain problematic, because it 'does not give back' to the land. Moreover, recent examples of infrastructure failure show that no amount of good planning can prevent environmental disasters from happening (Sandlos and Keeling 2016; Herz 2020), and "slow violence" (Nixon 2011) will remain a problem.

Repression of culture and language, intergenerational blame, and lateral violence caused by residential schools and dispossession of land are felt acutely by young adults who want to break the cycle of abuse. Reviving Indigenous languages is seen as one of the pivotal factors to re-indigenize life in the North, but doing so can be difficult in places heavily impacted by extractive colonialism. Language itself can be seen as ritual: Just as with cer-emony, it repeats its elements and varies them only according to rules defined by its grammar. Han (2019) proposes that rituals and ceremony are diametri-cally opposed to the industrial production logic. Following this argument, colonial extractivism could be seen as why it is First Nations languages and ceremonies which are most endangered in communities impacted heavily by the extractive industry, such as Mayo. Responsibility, reciprocity, and humil-ity are important Indigenous values, and strengthened by ceremony, art, and play. Indigenous youth are motivated by festivals, community activities, dance groups, hand games, and outdoor/cultural camps, which enable con-nectivity and relationship-building—essential aspects of environmental repossession (Big-Canoe and Richmond 2014) and reinhabitation (Trinidad 2012)—building up strong future leaders and resilient peoples.

In the Yukon, the extracted minerals lie beneath a vast landscape of hills, mountains and forests, interlaced by many rivers and lakes: home to many bears, moose, caribou, a variety of birds and insects, fish and other mam-mals. Respecting and recognizing these other-than-human persons as ances-tors, friends and teachers is one of the central aspects of a cosmology that values other-than-human beings including bodies of water and mountains not just as resources but as sentient relations (Ingold 2000; Cruikshank 2005, 2012). Maintaining a positive connection to 'all their [Indigenous] relations' (LaDuke 1999; Talaga 2020) is repeatedly emphasized as crucial by the young adults who participated in this study. Diverse opinions on many issues were expressed, but most participants stress that mining needs to be done respectfully—in a good way—by proper involvement of the local and First Nation population, values and protocols, including care for the environment and remediation upon mine closure. The systemic issues caused by mining colonialism or extractive imperialism/violence (Petras and Veltmeyer 2014; Sehlin MacNeil 2017) also need to be addressed—from individual behaviour modification to a change in mining companies' (speech) culture, to institutional change within authorities such as the police and social services. Systemic inequalities, rooted deeply within governance, health care and educational institutions, call for more than just reform; they also call for decolonizing state structures and indigenizing community life—and for respecting rights and working towards justice on a multitude of scales. The young adults who participated in this study have shown much passion for change, an acute capacity to identify problems, and offered fresh views and solutions. The findings of this study confirm that the impacts of mining are highly differential, cumulative (both additive and synergistic), contradictory, and sometimes hidden. They suggest that the long-term as well as the short-term impacts are neither solely positive nor negative, and that a wide range of phenomena are structurally related to each other when seen through the lens of extractivism. Finally, it is especially Indigenous youth who continue to carry many of the burdens associated with a colonial, extractivist past and present—the same people who continue to 'strive to survive and keep their culture and language alive.'

Acknowledgements

A heartfelt Mussi Cho to everyone who shared their stories with me, to my reviewers, as well as to the First Nation of Nacho Nyäk Dun and all other Yukon First Nations who welcomed me and my colleagues and supported this research on their traditional territories. This research was conducted as part of the ReSDA/SSHRC-funded project "LACE—Labour Mobility and Community Participation in the Extractive Industry. Case Study in the Yukon Territory" (2014–2019, PI Gertrude Saxinger and co-PI Chris Southcott), and the "Arctic Youth and Sustainable Futures" project (PI Joan Nymand Larsen; funded by Nordic Council of Ministers' Arctic Cooperation Programme and the Stefansson Arctic Institute).

Notes

1 The FN NND traditional territory is 162,456 km² large and expands into the Northwest Territories. Approximately 3% are designated 'Settlement Land' (Category A and B). The FNNND has ownership of surface and subsurface rights on Category A lands. The town of Mayo where about half of FN NND citizens live today, was established in 1903 following the finding of rich silver or more precisely galena ore veins in the region (FN NND Elders et al. 2019). Ninety years after the establishment of Mayo, in 1993, the FN NND signed their self-government agreement (Indian and Northern Affairs Canada 1993). Today, mining proponents are required to abide by the processes put in place by Yukon First Nations (see for example FN NND 2008a, 2008b, 2008c) and the Yukon state. Among the requirements are Impact Benefit Agreements, which often include provisions for local employment and supporting youth education—as well as

assessment by the Yukon Environmental and Socio-economic Assessment Board (YESAB 2020).

2 A community-based participatory research approach, including a focus living in a good way counters deficit framing—such as a focus on Indigenous Peoples being 'at risk' or particularly vulnerable, and victimry (Roe et al. 2012; Tsinnajinnie et al. 2019)—and fosters a critical "eye opening" experience for both the researcher and participants (Cahill et al. 2008). Lindroth and Sinevaara-Niskanen (2018) emphasize if Indigenous Peoples are particularly vulnerable, this vulnerability stems from historic and contemporary forms of oppression and structural challenges.

3 During 18 months of *Yo-Yo* fieldwork (Wulff 2002), I co-developed science communication products such as the 'Mobile Workers Guide—Fly-In/Fly-Out & Rotational Shift Work in Mining, Yukon Experiences' (Saxinger and Gartler 2017), the film "Mining on First Nation Land—The First Nation of Na-Cho Nyäk Dun in Mayo/Yukon Territory" (Saxinger and Gartler 2017), and 'Dän Hùnày—Our People's Story. First Nation of Nacho Nyäk Dun Elders' Memories and Opinions on Mining (FN NND Elders with Gartler et al. 2019).

4 A Youth Councillor's job is to voice Youth's concerns to Chief and Council, to bring opportunities to Youth within the First Nation, to build leadership skills and create so-called Northern Tutchone champions. (Interview with a Youth Councillor, 6 March 2017)

5 These participants were chosen due to: (1) their willingness to participate in the study; (2) their active leadership positions in respect to youth wellbeing; and (3) the relationship with the researcher (for the benefits of close relationships between researchers and participants see Ulturgasheva et al. 2015).

6 Two-Spirit, lesbian, gay, bisexual, transgender, queer, questioning, intersex and asexual (see NIIMMIWG 2020, p. 40).

7 For more information see the Websites http://protectpeel.ca/ and https://www.facebook.com/youthofthepeel/.

8 Since the interview the groups' name has changed to "Our Voices".

9 Part of the rap song lyrics of 'War Cry for Culture' by Driven to Change (2018), a song that expresses the thoughts of Yukon Indigenous Youth on the education system and colonialism (transcript by Susanna Gartler).

References

Acosta, A. (2013) 'Extractivism and neo-extractivism: Two sides of the same curse', in M. Lang, D. Mokrani, P. G. o. A. t. Development (eds), *Beyond development. Alternative visions from Latin America*, Luxembourg/The Netherlands: Transnational Institute Luxembourg – Rosa Foundation, pp. 61–87.

Assembly of First Nations (2021) 'National Chief RoseAnne Archibald Seeks Urgent Action from Federal Government after Release of Tk'emlúps te Secwèpemc Report on Unmarked Graves.' Available at: https://www.afn.ca/category/news-media/latest-news/.

Barker, B., Goodman, A., DeBeck, K. (2017) 'Reclaiming Indigenous identities: Culture as strength against suicide among Indigenous youth in Canada', *Canadian Journal for Public Health*, 108(2), pp. 208–210. doi: 10.17269/CJPH.108.5754.

Bartolovich, C. (2019) 'The common, force, and the capitalocene', *Minnesota Review*, 93, pp. 111–125. doi:10.1215/00265667-7737339.

Bélanger, P. (2018) *Extraction empire: Undermining the systems, states, & scales of Canada's global resource empire*. Cambridge, Massachusetts: The MIT Press. doi: 10.1111/cag.12654.

Bessarab, D., Ng'Andu, B. (2010) 'Yarning about yarning as a legitimate method in Indigenous research', *International Journal of Critical Indigenous Studies*, 3(50). doi: 10.5204/ijcis.v3i1.57.

Bhattacharyya, J., M. Baptiste, D. Setah, R. William (2013) 'It's who we are. Locating cultural strength in relationship with the land', in Parkins, J.R. and Reed, M. G. (eds) *Social transformation in rural Canada. Community, cultures, and collective action*, Vancouver/Toronto: UBC Press, pp. 211–231 doi: 10.25336/P6DK6V.

Big-Canoe, K., Richmond, C. A. M. (2014) 'Anishinabe youth perceptions about community health: Toward environmental repossession', *Health & Place*, 26, pp. 127–135. doi: 10.1016/j.healthplace.2013.12.013.

Bombay, A., Matheson, K., Anisman, H. (2014) *Origins of lateral violence in Aboriginal communities. A preliminary study of student-student abuse in residential schools*. Ottawa: Aboriginal Healing Foundation. Available at: http://www.ahf.ca/downloads/lateral-violence-english.pdf (Accessed: December 23 2020).

Burke, B. J. (2012) 'Transforming power in Amazonian extractivism: Historical exploitation, contemporary "fair trade", and new possibilities for Indigenous cooperatives and conservation', *Journal of Political Ecology*, 19(1), pp. 114–126. doi:10.2458/v19i1.21720.

Bush, E. and Lemmen, D.S. (eds) (2019) *Canada's changing climate report*. Government of Canada, Ottawa, ON.

Cahill, C., Rios-Moore, I., Threatts, T. (2008) 'Different eyes/open eyes community-based participatory action research', in M. Fine and J. Cammarota (eds), *Revolutionizing education: Youth participatory action research in motion, Critical youth studies*, New York/London: Routledge, Taylor & Francis Group, pp. 89–124. doi: 10.4324/9780203932100.

CBC News (2017) *Yukon First Nations woman receives award for tackling lateral violence*, CBC News, 8 Dec. Available at: https://www.cbc.ca/news/canada/north/teagyn-vallevand-award-samara-lateral-kindness-1.4440709 (Accessed: December 23 2020).

Chilisa, B. (2012) *Indigenous research methodologies*. Thousand Oaks, CA: Sage.

Clark, N. (2016) 'Red intersectionality and violence-informed witnessing praxis with Indigenous girls', *Girlhood Studies*, 9, pp. 46–64. doi: 10.3167/ghs.2016.090205.

Coates, K., Morrison, W. R. (2005) *Land of the midnight sun: A history of the Yukon* (2nd ed.). Montreal: McGill-Queen's University Press.

Council of Yukon First Nations (2020) '*Yukon First Nation climate action fellowship*'. Available at: https://cyfn.ca/climate-action-fellowship-targets-young-yukon-first-nations/ (Accessed: December 23 2020).

Cruikshank, J. (2005) *Do glaciers listen? Local knowledge, colonial encounters, and social imagination*. Vancouver: University of British Columbia Press. doi: 10.5038/2162-4593.11.1.9.

Cruikshank, J. (2012) 'Are glaciers 'good to think with? Recognising Indigenous environmental knowledge', *Anthropological Forum*, 22(3), pp. 239–250. doi: 10.1080/00664677.2012.707972.

Davison, C. M., Hawe, P. (2012) 'All that glitters: Diamond mining and Tłįchǫ youth in Behchokǫ, Northwest Territories', *ARCTIC*, 65(2), pp. 214–228. doi: 10.14430/arctic4202.

Demientieff, L. M. (2017) Deg Xit'an Athabascan conversations on wellness: A qualitative study exploring the radical possibilities of relationships. Doctoral Thesis, University of Utah.

Dhillon, J. (2017) *Prairie rising: Indigenous youth, decolonization, and the politics of intervention.* Toronto: University of Toronto Press. doi: 10.1353/gpq.2019.0022.

Driven to Change (2018) *War cry for culture. (Song) Council of Yukon First Nations.* Available at: https://www.youtube.com/watch?v=xUkiw2DnWgk (Accessed: December 23 2020).

First Nation of Nacho Nyäk Dun (2008a) '*Co-op engagement flowchart*', Mayo: First Nation of Nacho Nyäk Dun.

First Nation of Nacho Nyäk Dun (2008b) '*Cooperative engagement process for economic activities proposed in the Traditional Territory of the First Nation of Na Cho Nyak Dun*', Mayo: First Nation of Nacho Nyäk Dun.

First Nation of Nacho Nyäk Dun (2008c) '*Guiding principles towards best practices codes for mineral interests within First Nation of Na-Cho Nyak Dun Traditional Territory*', Mayo: First Nation of Nacho Nyäk Dun.

First Nation of Nacho Nyäk Dun Elders with Gartler, S., Hogan, J., Saxinger, G. (2019) 'Dän Hùnày – Our people's story'. *First Nation of Nacho Nyäk Dun Elders' Memories and opinions of mining.* Mayo: Yukon First Nation of Nacho Nyäk Dun/ Yukon College/ReSDA (Resources and Sustainable Development in the Arctic).

Gartler, S. (2018) 'One word, many worlds: The multivocality of "subsistence"', *Alaska Journal of Anthropology*, *16*(2), pp. 49–63. Available at: https://www.alaskaanthropology.org/wp-content/uploads/2021/02/Gartler_forweb.pdf.

Gartler, S., Kuklina, V., Schweitzer, P. (2019) 'Culture and sustainability', in J. K. Graybill & A. N. Petrov (eds), *Arctic sustainability, key methodologies and knowledge domains. A synthesis of knowledge I.* Abingdon, Oxon/New York: Routledge, pp. 61–86.

Gibson, G., Klinck, J. (2005) 'Canada's resilient North: The impact of mining on Aboriginal communities', *Pimatisiwin: A Journal of Aboriginal and Indigenous Community Health*, *3*(1), pp. 115–140.

Gignac, J. (2019) '*Blake Shaá'koon Lepine is part of a legion of artists working to change Indigenous art. The local artist has an exhibit at Arts Underground*', Yukon News, 3 September. Available at: https://www.yukon-news.com/entertainment/blake-shakoon-lepine-is-part-of-a-legion-of-artists-working-to-change-Indigenous-art/ (Accessed: December 23 2020).

Goldfarb, R. J., Hart, J. R. C., Miller, M. L., Miller, L. D., Farmer, L. G., Groves, D. I. (2000) The Tintina Gold Belt – A global perspective, in Tucker, T.L. and Smith M.T. (eds) *The Tintina Gold Belt: Concepts, exploration, and discoveries: Special vol. 2*, Vancouver, BC: British Columbia/ Yukon Chamber of Mines, pp. 5–34.

Han, B.-C. (2019) *Vom verschwinden der Rituale: Eine Topologie der Gegenwart.* Berlin: Ullstein.

Harvey, D. (2003) *The new imperialism.* Oxford: Oxford University Press.

Harvey, D. (2006) 'Neo-liberalism as creative destruction', *Geografiska Annaler: Series B, Human Geography*, *88*(2), pp. 145–158. doi: 10.1111/j.0435-3684.2006. 00211.x.

Harvey, T. (2007) 'Challenging the planet: A call from the Arctic to face the heat on climate change', in S. Cullis-Suzuki, K. Frederickson, A. Kayssi, C. Mackenzie & A. D. Cohen (eds), *Notes from Canada's young activists: A generation stands up for change*, pp. xii–226. Vancouver, Toronto, Berkeley: Greystone Books.

Hatala, A. R., T. Pearl, K. Bird-Naytowhow, A. Judge, E. Sjoblom, L. Liebenberg (2017) '"I have strong hopes for the future": Time orientations and resilience among Canadian Indigenous youth', *Qualitative Health Research*, 27(9), pp. 1330–1344. doi: 10.1177/1049732317712489.

Herz, N. (2020) '*As Arctic warming accelerates, permafrost thaw hits Red Dog mine with $20 million bill*', Anchorage: Alaska Public Media, 1 September. Available at: https://www.alaskapublic.org/2020/09/01/as-arctic-warming-accelerates-permafrost-thaw-hits-red-dog-mine-with-20-million-bill/ (Accessed: December 23 2020).

Holden, W., Nadeau, K., Jacobson, R. D. (2011) 'Exemplifying accumulation by dispossession: Mining and Indigenous peoples in the Philippines', *Geografiska Annaler: Series B, Human Geography*, 93(2), pp. 141–161. doi: 10.1111/j.1468-0467.2011.00366.x.

Holen, D., Gerkey, D., Høydahl, E., Natcher, D., Reinhardt Nielsen, M., Poppel, B., Aslakse, J. (2015) 'Interdependency of subsistence and market economies in the Arctic', in S. Glomsrod, G. Duhaime, J. Aslaksen (eds), *The Economy of the North*, Oslo: Statistics Norway, pp. 89–126.

Huskey, L., Southcott, C. (2016) '"That's where my money goes": Resource production and financial flows in the Yukon economy', *The Polar Journal*, 6(1), pp. 11–29. doi: 10.1080/2154896X.2016.1171002.

Indian and Northern Affairs Canada (1993) *The First Nation of Na-Cho Nyäk Dun self-government agreement*. Ottawa: Ministry of Indian Affairs and Northern Development.

Ingold, T. (2000) *The perception of the environment: Essays on livelihood, dwelling and skill*. London and New York: Routledge.

IRC – Inuvialuit Regional Corporation (2016) 'Inuvialuit on the frontline of climate change: Development of a regional climate change adaptation strategy'. Available at: https://irc.inuvialuit.com/system/files/Inuvialuit%20on%20the%20Frontline%20of%20Climate%20Change-Final-Feb2018%20%20%28SMALL%29.pdf.

Klein, N. (2011) 'Capitalism vs. the climate'. *The Nation*, 9 November. Available at: https://www.thenation.com/article/archive/capitalism-vs-climate/ (Accessed: December 23 2020).

Kral, M. J. (2012) 'Postcolonial suicide among Inuit in Arctic Canada', *Culture Medicine and Psychiatry*, 36(2), pp. 306–325. doi: 10.1007/s11013-012-9253-3.

Kral, M. J. (2013) '"The weight on our shoulders is too much, and we are falling": Suicide among Inuit male youth in Nunavut, Canada', *Medical Anthropology Quarterly*, 27(1), pp. 63–83. doi: 10.1111/maq.12016.

Kral, M. J., Salusky, I., Inuksuk, P., Angutimarik, L., Tulugardjuk, N. (2014) 'Tunngajuq: Stress and resilience among Inuit youth in Nunavut, Canada', *Transcultural Psychiatry*, 51(5), pp. 673–692. doi: 10.1177/1363461514533001.

Kulchyski, P. K. (2005) *Like the sound of a drum: Aboriginal cultural politics in Denendeh and Nunavut*. Winnipeg: University of Manitoba Press.

LaDuke, W. (1999) *All our relations: Native struggles for land and life*. Chicago: Haymarket Books.

Laidler, G., Ford, J., Gough, W., Ikummaq, T., Gagnon, A., Kowal, S., Qrunnut, K., Irngaut, C. (2009) 'Travelling and hunting in a changing Arctic: Assessing Inuit vulnerability to sea ice change in Igloolik, Nunavut', *An Interdisciplinary, International Journal Devoted to the Description, Causes and Implications of Climatic Change*, 94, pp. 363–397. doi: 10.1007/s10584-008-9512-z.

Larsen, J. N., Huskey, L. (2020) 'Sustainable economies in the Arctic', in J. K. Graybill & A. N. Petrov (eds), *Arctic sustainability, key methodologies and knowledge domains: A synthesis of knowledge I.* Abingdon, Oxon/New York: Routledge, pp. 23–42.

Lavallee, L., Poole, J. (2010) 'Beyond recovery: Colonization, health and healing for Indigenous people in Canada', *International Journal of Mental Health and Addiction*, 8, pp. 271–281. doi: 10.1007/s11469-009-9239-8.

Lindroth M., Sinevaara-Niskanen H. (eds) (2018) 'The neoliberal embrace of resilient Indigeneity', in *Global politics and its violent care for Indigeneity*. Cham: Palgrave Macmillan. Springer International Publishing, pp. 79–102.

Liu, J. (2019) 'First Nations, first-hand art: A look at Yukon artists rebuilding their culture', *CBC News*, 27 June. Available at: https://www.cbc.ca/news/canada/north/yukon-first-nations-art-revitalization-1.5189360 (Accessed: December 23 2020).

Logie, C., Lys, C., Mackay, K., MacNeill, N., Pauchulo, A., Yasseen, A. (2019) 'Syndemic factors associated with safer sex efficacy among northern and Indigenous adolescents in Arctic Canada', *Official Journal of the International Society of Behavioral Medicine*, 26, pp. 449–453. doi:10.1007/s12529-019-09797-0.

Malm, A., & Hornborg, A. (2014) 'The geology of mankind? A critique of the Anthropocene narrative', *The Anthropocene Review*, 1, pp. 62–69. doi: 10.1177/2053019613516291.

Marcus, D. S. (2016) 'Indigenous hermeneutics through ceremony: Song, language, and dance', in D. L. Madsen (ed.), *The Routledge companion to Native American literature*. Abingdon, Oxon/New York: Routledge, pp. 440–452.

McClellan, C., with Birckel, L., Bringhurst, R., Fall, J. A., McCarthy, C., Sheppard, J. R. (1987) *Part of the land, part of the water: A history of the Yukon Indians*. Vancouver: Douglas & McIntyre.

Moore, J. W. (2015) *Capitalism in the web of life: Ecology and the accumulation of capital* (1 publ. ed.). London: Verso.

Nadasdy, P. (2017) *Sovereignty's entailments: First Nation state formation in the Yukon*. Toronto, Buffalo, London: University of Toronto Press.

Natcher, D. C., Davis, S. (2007) 'Rethinking devolution: Challenges for Aboriginal resource management in the Yukon Territory', *Society and Natural Resources*, 20(3), pp. 271–279. doi: 10.1080/08941920601117405.

National Inquiry into Missing and Murdered Indigenous Women and Girls (NIIMMIWG) (2020) *Reclaiming power and place: The final report of the national inquiry into missing and murdered Indigenous women and girls (Vol. 1A)*. Available at: https://www.mmiwg-ffada.ca/ (Accessed: December 23 2020).

Natural Resources Canada (2019) *Minerals sector employment. Information bulletin.* Available at: https://www.nrcan.gc.ca/science-data/science-research/earth-sciences/earth-sciences-resources/earth-sciences-federal-programs/minerals-sector-employment/16739#fn10-rf (Accessed: December 23 2020).

Nelson, M., Natcher, D. C., Hickey, C. G. (2005) 'Social and economic barriers to subsistence harvesting in a Northern Alberta Aboriginal community', *Anthropologica*, 47(2), pp. 289–301. Available at: https://www.jstor.org/stable/25606241.

Nixon, R. (2011) *Slow violence and the environmentalism of the poor*. Cambridge: Harvard University Press.

Parlee, B., Marlowe, M. (2002) *Community-based monitoring: Final report for the west Kitikmeot Slave Study Society*. Yellowknife: Lutsel K'e Dene First Nation, West Kitikmeot Slave Study Society.

Petras, J. F., Veltmeyer, H. (eds) (2014) *Extractive imperialism in the Americas: Capitalism's new frontier*. Leiden, Boston: Brill. doi: 10.1163/9789004268869.

Petrasek MacDonald, J., Cunsolo Willox, A., Ford, J. D., Shiwak, I.,Wood, M. (2015) 'Protective factors for mental health and wellbeing in a changing climate: Perspectives from Inuit Youth in Nunatsiavut, Labrador', *Social Science & Medicine, 141*, pp. 133–141. doi: 10.1016/j.socscimed.2015.07.017.

Rasmus, S. M. (2014) 'Indigenizing CBPR: Evaluation of a community-based and participatory research process implementation of the Elluam Tungiinun (towards wellness) program in Alaska', *American Journal of Community Psychology, 54*(1–2), pp. 170–179. doi:10.1007/s10464-014-9653-3.

Rasmus, S. M., Allen, J., Ford, T. (2014) '"Where I have to learn the ways how to live:" Youth resilience in a Yup'ik village in Alaska', *Transcultural Psychiatry, 51*, pp. 713–734. doi:10.1177/1363461514532512.

Regan, P. (2010) *Unsettling the settler within: Indian residential schools, truth telling, and reconciliation in Canada*. Vancouver: UBC Press.

Roe, Y. L., Zeitz, C. J., Fredericks, B. (2012) 'Study protocol: Establishing good relationships between patients and health care providers while providing cardiac care. Exploring how patient-cinician engagement contributes to health disparities between Indigenous and non-indigenous Australians in South Australia', *BMC Health Services Research, 12*, pp. 397–397. doi:10.1186/1472-6963-12-397.

Russell-Mundine, G. (2012) 'Reflexivity in Indigenous research: Reframing and decolonising research?' *Journal of Hospitality and Tourism Management, 19*, pp. 85–90. doi: 10.1017/jht.2012.8.

Sandlos, J., Keeling, A. (2016) 'Toxic legacies, slow violence, and environmental injustice at Giant Mine, Northwest Territories', *Northern Review, 42*, pp. 7–21. doi: 10.22584/nr42.2016.002.

Saxinger, G. (2018) 'Community based participatory research as a long-term process: Reflections on becoming partners in understanding social dimensions of Mining in the Yukon (Essay)', *Northern Review, 47*, pp. 187–206. doi: 10.22584/nr47.2018.009.

Saxinger, G., Gartler, S. (2017) *The mobile workers guide. Fly-in/Fly-out & rotational shift work in mining. Yukon Experiences*. Whitehorse: Yukon First Nation of Nacho Nyäk Dun/Yukon College/ReSDA (Resources and Sustainable Development in the Arctic).

Saxinger, G., Gebauer R., Oschmann J., and Gartler, S. (2017) Film 'Mining on First Nation Land: The First Nation of Nacho Nyäk Dun', 13 min., produced by the First Nation of Nacho Nyäk Dun. Available at: https://youtu.be/u4UXywmkoqM.

Saxinger, G., Öfner, E., Shakirova, E., Ivanova, M., Yakovlev, M., Gareyev, E. (2016) 'Ready to go! The next generation of mobile highly skilled workforce in the Russian petroleum industry', *The Extractive Industries and Society, 3*, pp. 627–639. doi: 10.1016/j.exis.2016.06.005.

Sehlin MacNeil, K. (2017) Extractive violence on Indigenous country: Sami and Aboriginal views on conflicts and power relations with extractive industries. Umeå universitet, Centrum för samisk forskning (CeSam).

Smith, L. T. (2012) *Decolonizing methodologies: Research and Indigenous peoples* (2nd ed.). London and New York: Zed Books and University of Otago Press.

Talaga, T. (2020) *All our relations. Indigenous trauma in the shadow of colonialism*. Toronto: Scribe.

Trinidad, A. M. O. (2012) 'Critical Indigenous pedagogy of place: A framework to Indigenize a youth food justice movement', *Journal of Indigenous Social Development, 1*(1), pp. 1–17. doi: 10.1080/15313204.2014.903136.

Tsinnajinnie, L. R. R., Minthorn, R. S. Z., Lee, T. S. (2019) 'K'é and Tdayp-tday-gaw: Embodying Indigenous relationality in research methods', in Sweeney Windchief and Timothy San Pedro (eds), *Applying Indigenous research methods: Storying with peoples and communities, Indigenous and decolonizing studies in education*, New York: Routledge, pp. 37–55. doi: 10.4324/9781315169811-3.

Ulturgasheva, O., Rasmus, S., Morrow, P. (2015) 'Collapsing the distance: Indigenous-youth engagement in a circumpolar study of youth resilience', *Arctic Anthropology*, *52*(1), pp. 60–70. doi: 10.3368/aa.52.1.60.

Ulturgasheva, O., Rasmus, S., Wexler, L., Nystad, K., Kral, M. (2014) 'Arctic Indigenous youth resilience and vulnerability: Comparative analysis of adolescent experiences across five circumpolar communities', *Transcultural Psychiatry*, *51*(5), pp. 735–756. doi: 10.1177/1363461514547120.

Willow, A. J. (2016) 'Indigenous ExtrACTIVISM in boreal Canada: Colonial legacies, contemporary struggles and sovereign futures', *Humanities*, *5*(55), pp. 1–15. doi: 10.3390/h5030055.

Winnemucca, S. (1883/1969) *Life among the Piutes: Their wrongs and claims*. Bishop, CA: Sierra Media Inc.

Winton, A., Hogan, J. (2015) '"It's just natural": First Nation family history and the Keno Hill silver mine in the Yukon Territory', in A. Keeling, J. Sandlos (eds), *Mining and communities in Northern Canada: History, politics, and memory*, Calgary: University of Calgary Press, pp. 87–116.

Wulff, H. (2002) 'Yo-yo fieldwork: Mobility and time in a multi-local study of dance in Ireland', *Anthropological Journal on European Cultures*, *11*, pp. 117–136. Available at: https://www.jstor.org/stable/43234897.

YESAB (2020) *Yukon environmental and socio-economic assessment board website. The assessment process.* Available at: https://www.yesab.ca/the-assessment-process/ (Accessed: December 23 2020).

Interviews

- Interview with female member of FN NND Heritage and Culture Department (58 mins). September 19, 2014.
- Interview with female FN NND Youth councillor (33 mins). October 2, 2014, Mayo.
- Interview with male mine worker (36 mins). October 3, 2014, Mayo.
- Interview with male mine worker (21 mins). October 3, 2014, Mayo.
- Interview with male mine worker (27 mins). October 15, 2014, Mayo.
- Interview with male mine worker (29 mins). October 21, 2014, Wolverine Mine Camp.
- Interview with female mine worker (37 mins). October 22, 2014, Wolverine Mine Camp.
- Interview with male mine worker (51 mins). June 28, 2015, Mayo.
- Interview with male youth (1 hour 9 mins). February 16, 2017. Whitehorse, Yukon Inn restaurant.
- Interview with female FN NND Youth councillor (55 mins). March 6, 2017, Mayo.
- Interview with female youth (1 hour 54 mins). November 8, 2018. Mayo, Yukon College campus.
- Interview with one female and one male youth, (41 mins). November 21, 2018. Mayo, Yukon College campus.
- Interview with FN NND Community Education Liaison Coordinator (41 mins). January 21, 2019, Mayo at J.V. Clark School.

Part III

Regulating youth's paths to independence

7 Youth law, policies and their implementation in the Russian Arctic

*Aytalina Ivanova, Tatyana Oglezneva and
Florian Stammler*

Introduction

There is a high demand for youth in the Russian Arctic as the territories making up the region should be strategic drivers of development in the country at large (Fondahl et al. 2020, pp. 195–196). Yet the region is losing population and suffers from a lack of qualified labour. In this chapter our starting point is that a low working-age population remains the most significant problem where development of the Russian Arctic is concerned. Post-Soviet state policy has not yet solved this problem. Retaining young people who were born in the region is particularly important for decreasing outmigration (Volgin et al. 2018, pp. 44–45). In this respect, the Russian Arctic has a combination of material and moral incentives for attracting and retaining youth. These include a fairly well-developed system of cities and large urban settlements which have survived from the Soviet period of northern development (Zamiatina 2020). On the other hand, the Soviet industrial heritage is not enough to keep young qualified people living in Arctic cities (see Bolotova, this volume).

This chapter first outlines the regulatory framework for Russian youth policy in the Arctic in light of the current demographic, economic and social challenges facing the region. We adhere to the administrative and legal structure of the Russian Federation, looking first at federal and then at regional laws and policies. An analytical section on implementation then draws on case studies in the cities of Neryungri (Yakutia), Novy Urengoy (Yamal), and Kirovsk/Apatity (Murmansk Region). Based on document analysis and on-site fieldwork visits, we analyse the local implementation of particular policies and programmes and their acceptance among local young people.

Public debates and discussions about policy and regulation reveal that there are two main tendencies when it comes to assessing human capital in the Arctic: one group in the debate maintains that Russian Arctic cities are overpopulated as a legacy of Soviet settlement policies (Hill and Gaddy 2003); the other, whose argument we support, contends that one cannot develop the Arctic without people and that incentives should be created for people to stay or come to the North (Parente et al. 2012 and in this volume: Adams et al.; Bolotova; Komu and Adams;

DOI: 10.4324/9781003110019-11

Simakova et al.). Since the end of the Soviet Union, the Russian Arctic has experienced a significant outflow of population, although this has not occurred evenly across all parts of the region (Heleniak 2009). For example, over the past 20 years the population in the 13 Arctic municipalities of the Republic of Sakha (Yakutia) has decreased by more than half, from 148,000 to 68,000 persons. In comparison, the decline in the permanent population of the Russian Arctic at large has been less drastic, with a decrease from 2,400,000 to 2,371,000 recorded for 2014 and 2017 (Mikhailovskaia 2018a).

Many regions in the Russian Arctic require a larger labour force. The need for young trained specialists in the North annually is 25,000 people (Volgin et al. 2018, p. 46). The outflow of graduates, the region's limited educational opportunities and the need to train young specialists to meet the requirements of the companies working in the Arctic are acute concerns. Accordingly, there is an urgent need to create a set of measures to attract and retain specialists from other regions of Russia. In recent years, this question has received a great deal of attention among various stakeholders in Russia, which can be seen from an increase in the range of conferences organized on this topic. Participants in these forums have come up with important suggestions and avenues for development, several of which are illustrated below.

One recent conference resolution stated:

> [in order] to provide the Arctic with young professional personnel and to retain the population of the northern regions, it is currently important to determine the life strategies of modern youth in the Arctic and to create suitable conditions for their professional and social self-realization.
>
> (Assotsiatsiia Poliarnikov 2018, p. 5)

Many activists and decision-makers at the nationwide meeting agreed that the first and main condition for the consolidation of youth in the Arctic is to improve infrastructure, transport and further cultural and social development. One idea was to launch a broad information campaign on the benefits of living in the North.

Another recent Arctic meeting in Russia featured a special "day of youth" (Assotsiatsiia Poliarnikov 2019), a forum where young people from the Arctic shared their experiences of participating in various Russia-wide youth competitions. This form of collective action is popular throughout the country and provides people with avenues for realizing their dreams and plans. The occasions offer young people the opportunity to make new contacts, as well as inspiration in the form of new ideas for solving problems of their home, which many call "*malaya rodina*" ("small homeland", meaning one's specific place of birth or living, as distinct from Russia as a whole). According to the Russian Ministry for Far Eastern and Arctic Development, business in the Arctic must be stimulated by preferential credit rates for the import of goods as a means to offset the increased expenses of delivering supplies to the North. The new law on business support for the Arctic embodies this approach (Russian Federation 2020; Trutnev and Kozlov 2020).

At yet another meeting in March 2017 (Dolgova et al. 2017), young professionals came up with their own proposals for Arctic development. They considered that the key goal of the Arctic Development Strategy should be to ensure human wellbeing in a stable and safe Arctic. Furthermore, they emphasized that any youth strategy for the Arctic should prioritize a balance between intensive innovative socio-economic development and sustainable development.

Notwithstanding such numerous meetings and an increased interest in the Arctic, we are of the view that Russian youth policy does not yet take into account the specifics of the northern territories and that the relevant laws in the field of youth policy are still insufficiently integrated (Federative Council 2009, p. 3).

To promote the interests of young people, a national Youth Chamber was established in 2001 under the State Duma, later, in 2011, becoming the Russian Youth Parliament (Mikhailovskaia 2018b). Thereafter, Arctic regions duplicated this system, establishing regional youth parliaments of their own. For example, in our case regions, the youth chambers became youth parliaments in 2016 in Yamal (Zakonodatel'noe Sobranie YNAO 2016), in 2017 in Murmansk Region (Severpost 2017) and in 2019 in Yakutia (Press-sluzhba Il Tumena 2019). Such forums have started working within the last ten years in almost all Arctic regions. The main goal of the new platform is to improve dialogue between the Arctic regions of the Russian Federation, promote regulation of the rights and legitimate interests of young people in those regions, develop projects aimed at stopping the outflow of young people and from the Arctic and create a positive image of Russia's activities there (Mikhailovskaia 2018b). Other aims include systematic work to highlight the possibilities of a decent life, work and self-realization for students in the harsh conditions of the North, for which young people need to be made aware of opportunities for professional growth and development. This mechanism makes it is possible to develop a new strategy, different from that practiced in Soviet times, to attract young people to live and work in the Arctic.

Russia's Arctic youth policy

In the absence of a Youth Act, youth in Russia are covered by a system of social obligations and guarantees that applies across the entire country. These are outlined in the Russian youth policy programme (Government of the Russian Federation 2014), as well as different government-funded programmes, for example the one focusing on housing for young families (Government of the Russian Federation 2016). However, as these programmes do not consider the particular conditions in the North, another set of privileges has been established for all Russian citizens and even foreigners with the right to work and reside in the Arctic (Konsul'tantPlius 2017). This is the so-called "regional coefficient", a bonus added to wages that are paid in areas officially belonging to the Far North of Russia. Such salary top-ups for people working in the North were first introduced during the Soviet Union after Stalin. The system is based on the principle that the harsher the climatic conditions of the area are for work, the higher the top-up should be (State archive Murmansk 2012). It continues to play an important role for the demography of the Russian North as a whole.

Russia is organized as a federal state that is divided into different regional administrative entities (known as *subekty*), which have the right to pass their own legislation in areas of regional relevance provided this does not conflict with federal law. Given this competence, seven Arctic units of the Russian Federation have passed legislation entitling all young employees of state institutions who are under 30 years of age to receive the full salary bonus from the first day of work. The regions are the Republic of Sakha (Yakutia), the Republic of Komi, the Krasnoyarsk Territory, the Arkhangelsk and Murmansk Regions and the Yamal-Nenets and Chukotka Autonomous Districts. Youth law in general is a field where Russian regions have seen fit to exercise their legal competence: even though there is no such law at the federal level, all of the nine Arctic regions of the Russian Federation have enacted their own youth legislation regulating social relations in the field of youth policy, albeit with varying degrees of detail and effectiveness. These regions are the Republics of Karelia, Sakha (Yakutia), and Komi, the Krasnoyarsk Territory, the Arkhangelsk and Murmansk Regions and the Yamal-Nenets, Nenets and Chukotka Autonomous Districts. Seven of the Arctic regions (the exceptions being Murmansk and Chukotka) have enacted laws on youth and youth policy, some of which are described in detail below. These laws determine the general objectives and principles of youth policy in the respective regions. The stated goals include the following:

- establishment of methods of informational, scientific, methodological and personnel support in the field of youth policy;
- organizational and legal support of the main directions of youth policy;
- organization of youth policy management; development of mechanisms strengthening legal regulation and the uniformity of youth policy at the regional and municipal levels;
- youth policy financing;
- measures to support youth in the field of education;
- vocational guidance;
- labour and employment assistance;
- support of entrepreneurship; and
- support with housing for young families.

Several specific examples from the regions will serve to show the orientation of such youth laws.

In the Republic of Sakha (Yakutia) (Il Tumen 2018), organizations creating additional youth jobs may be granted tax benefits. Employers owned fully or partially by the regional authorities are required to create jobs for young professionals permanently residing in the republic, specifically persons up to

30 years of age with secondary vocational or higher education. Companies with fewer than 100 employees need to provide at least 1 per cent (meaning a minimum one job) of such jobs. Moreover, at the birth of their first child (or adoption of a child under the age of three months), young parents in the republic have the right to receive a one-time allowance for the purchase of the child's first needs and food (today RUB 7048) (Ob okhrane sem'i, materinstva, ottsovstva i detstva 2008).

In the Yamal-Nenets Autonomous District (hereafter YNAO) (O molodezhnoi politike v Iamalo-Nenetskom avtonomnom okruge 2009), young specialists whose first principal place of employment falls under an employment contract in state and municipal institutions related to youth are provided with a one-time payment, as well as a monthly allowance paid for the first three years from the day of employment, while they are under the age of 31 years.

In all Russian regions, housing is considered a key area in which youth need state support. This being the case, in addition to the above-mentioned federal programme, regional governments develop and implement measures for subsidizing interest rates on housing loans, which attracts young citizens to work in the public sector in state and municipal institutions.

While this shows a significant "niche of agency" (Fondahl et al. 2019) for regions in the absence of applicable federal law, the main problem related to implementation is the lack of funding for the measures adopted. Furthermore, the conditions in the Arctic impose specific financial burdens on the regions stemming from the climatic and transport conditions, such as the difficulty of delivering fuel and food to remote areas within what is a very narrow seasonal window. In short, the public sector is more expensive to maintain in the Arctic region, whereby the degree to which Arctic territories can be developed depends greatly on budgetary security and on financial assistance from the federal treasury.

Article 41 of the Russian Labour Code gives businesses the right to take on increased financial obligations, for example in a collective agreement. Taking into account its financial and economic situation, a company can, for example, pay benefits to employees, create more favourable working conditions or make agreements offering terms going beyond the minimum established by law.

One such example is the Russian State Railway Company (RZD), which, in 2018, introduced amendments to the corporate salary system allowing the managers of the different regional railway divisions to switch over to a more favourable system of allocating salary top-ups to youth in the North (Truntseva 2018). The Northern and the East Siberian Division of RZD have opted for this system so far. The Northern Division considers the abolition of seniority as a prerequisite for obtaining the northern salary top-ups to be good motivation for young professionals. For example, a young person working in the Arkhangelsk subdivision of the northern railways will receive a bonus of 50 per cent on top of the base salary, just as his or her senior colleagues do. In the Yamal-Nenets Autonomous District, this figure is 80 per cent. In other words, a railway worker there is paid 80 per cent more for the same work than his or her colleague working in a region without northern salary top-ups. The introduction of such regulations has increased the competitiveness of railroad workers' wages in regional labour markets. This can stimulate an influx of young workers and specialists to the Russian Arctic. However, since the regulation is recent (2018), its effectiveness cannot yet be demonstrated in numerical terms.

Such measures show that not only regional authorities in Russia but also large corporate actors have opportunities to influence youth policy and regulation in the Arctic to their own advantage, creating incentives for youth. This indicates, as we have argued elsewhere (Stammler and Ivanova 2016; Fondahl et al. 2019), that there is more agency in Russia's Arctic regions than the image of a strong centralised federal state with a strong president might suggest.

Implementation of youth policy in Russian Arctic single-industry towns

In this section we show that opportunities to influence conditions of wellbeing for youth do not stop at the regional level in the Russian Arctic; cities and municipalities also have certain legal competences at their disposal. According to federal law, municipalities are responsible for working with children and youth (Russian Federation 2019). For the local governments of single-industry towns, this means an additional budgetary burden if they are to retain and attract youth. In our research project, we studied how four cities in this category used their competences: Neryungri (Republic of Sakha (Yakutia)), Novy Urengoy (Yamal-Nenets Autonomous District), Apatity and Kirovsk (Murmansk Region), thus adding to the ethnographic depth of other chapters in this volume focusing on the same cities (Bolotova, Simakova et al.).

Neryungri—"You decide what your city looks like"

The municipality of Neryungri adopted a five-year programme for implementing youth policy (2017–2021) (Administratsiia Neriungrinskogo Raiona 2016), the main aim of which is to improve conditions and guarantees for the economic and social wellbeing of young people. Neryungri's population, like that of many other single-industry towns in the northern and Arctic territories, is still decreasing slightly each year. In the first ten years of its existence, the city grew very rapidly, reaching 100,000 inhabitants in 1985. After the collapse of the Soviet Union, about half the population of the city left, but recently the decrease has slowed: between 2010 and 2019, the population of the city decreased from 61,747 people to 56,888 people. However, the proportion of young people in town is quite large (as much as 30 per cent), making the youth policy programme significant. The programme comprises the following measures:

- "Support for youth initiatives and the provision of socio-psychological support";

- "Patriotic education of youth"; and
- "Support for socially oriented non-profit organizations".

Another municipal five-year programme focuses on providing housing to young families, defined as those in which the spouses are under 35 years old. This is co-financed by all three levels of the Russian state (federal, regional, municipal) as well as non-state sources. At the beginning of 2016, 25 young families were registered for getting support from the programme. The five-year funding for this programme would provide the equivalent of approximately twelve apartments in Neryungri (Administratsiia Neriungrinskogo Raiona 2018a). If the funds are used for financing a mortgage or bank loan, more families can improve their housing situation with that money. An additional support measure has been in effect in the Neryungri district for pupils and students from low-income families (including large families) that pays compensation to cover the cost of travelling to a place of study within the district (Administratsiia Neriungrinskogo Raiona 2018c, 2018b).

Since Neryungri is a single-industry town, the main company in the city can also significantly influence the attractiveness of the city for its young inhabitants. Especially where municipal budgets are small, various forms of public–private co-financing are one way of increasing the effectiveness of youth programmes. The main coal mining company in Neryungri, Yakutugol, participates in the Yakutia-wide programme "Local personnel in the industry". Close cooperation has been established with the governments and employment centres of the rural municipalities. At career counselling meetings, company employees can inform local residents about working conditions, requirements for applicants, the level of salaries, as well as social guarantees and opportunities for professional self-realization.

In practice, while these measures have not stopped outmigration, we can still state, based on our fieldwork, that they have contributed to slowing the downward development in the city and the atmosphere of decay that Bolotova (this volume) identified in shrinking cities. Several examples will serve to demonstrate this: While Neryungri remains a single-industry town, its economic profile has been diversified and additional work opportunities for young people have been created. One would think that building housing is not a priority in a town where the population has shrunk by more than 30,000 people from its peak. However, fieldwork conversations show that the old Soviet housing is not attractive enough for young people, with some houses being dilapidated wooden barracks and others ill-maintained 1970s Soviet-style apartments. In our focus group discussion with young employees of both major coal mining companies, housing was among the most important topics mentioned in connection with their sense of a good life. In 2017, 400 new apartments were built by the municipality; during our fieldwork, we witnessed young people moving into these buildings (Figure 7.1).

Higher and other post-secondary educational opportunities are another important "plus" for this northern town, without which the outmigration

Figure 7.1 Young family moving in one of the new apartments in Neryungri, Churapchinskaia Street.

(Photo: F. Stammler, 2018)

might have been higher. These include the programmes at a technical institute which is a branch of Yakutsk North Eastern Federal University and educates specialists in high demand in town. According to Director Pavlov (2018), all graduates in the mining-related professions can find a job in town if they want. The institute also offers degrees in philology and economics. On a more vocationally oriented level, South Yakutian college educates technical specialists, who are also in high demand in the town. Retaining young educated people in town also creates a demand for more facilities to promote social wellbeing. The city tries to meet this need with various initiatives increasing a positive atmosphere, one being a citizen participation contest geared to upgrading public spaces such as parks, walkways and bus stops, with the slogan "you decide how your city looks like".

In civil society, churches are important actors, offering as they do activities for young families. One opportunity of note in Neryungri is the youth activity of the evangelical church, which has a very active youth pastor, Alexandr Proshchalgin. In his "laboratory of applied social technologies", as he calls it, he organizes a variety of youth activities. These include extreme sports activities, outdoor bonding between fathers and sons, motorsport activities, counselling for young families, as well as drug and alcohol prevention measures such as life orientation through lectures and activities. He counted that more than 1500 young people between the ages of 16 and 20 have already taken part in such measures, and is proud to announce that some have gone from being on the verge of a life of crime to becoming millionaires (Proshchalgin 2018).

Novy Urengoy—how much rootedness can gas money buy?

The city of Novy Urengoy has different starting conditions than most other Russian industrial towns. It advertises itself as the "gas capital of Russia", as the discovery of two of Gazprom's biggest deposits in the vicinity were the reason for its establishment. The population has increased slightly in the last ten years, from 104,000 to 118,000 inhabitants (Rosstat 2020). In this single-industry town reliant on gas revenues, making money has been an important goal among many inhabitants since Soviet times (Stammler 2010). An analysis of the situation of the town's youth shows that despite increased non-monetary incentives to make life there attractive in recent years, young people's tendency to focus solely on increased material wellbeing has remained unchanged since Soviet times: the idea is to make enough money while living in Novy Urengoy and then relocate to a more comfortable place. Thus, even decades after the end of the Soviet Union, the city retains the image of a transient residence and this cannot be ascribed solely to its Soviet heritage. As previous research has shown, even though "The City Became the Homeland for its Inhabitants, (...) Nobody is Planning to die here" (Stammler 2010, p. 33).

The municipal programme "Children and Youth. Development of Civil Society 2014–2020" aims at creating legal, economic and organizational conditions for young people's personal development and support for public youth associations in order to increase their social wellbeing (Administratsiia goroda Novyi Urengoi 2017). The programme is divided into two sections, "Social and personal development of youth and children's travelling for recreation during the holidays" and "Support of socially oriented non-profit organizations", with the latter able to get staff salaries from programme funds. The programme is administered by the Department for Youth and Public Affairs of the Administration of the City of Novy Urengoy. Several institutions targeting different groups of youth are financed from the department's budget. For example, the youth centre "Nord" has a focus on motorsports (go-karts, motocross bikes and snowmobiles) and organizes survival training outdoors, partially combined with what the organizers call "military patriotic education". While participants learn how to survive outdoors, they also learn how to assemble a Kalashnikov automatic rifle, get to test military gear and learn about the Russian army. It is worth noting that girls also actively take part in what is usually a male-dominated sphere. Some make it to regional or Russia-wide competitions, for example, in motocross or snowmobile racing. The city finances the building used by the centre, the salaries of the staff and some basic equipment. Members of "Nord" contribute by buying their own equipment, which they bring to the centre. In the workshops young people also learn to fix their motorbikes, go-karts and snowmobiles with instruction from trainers paid by the city. The centre has more than 200 active members, but many of the young people come with their parents (Vorobiev 2019) (Figure 7.2).

Another youth centre, "Optimist", is more oriented towards indoor activities with youth. It provides space for youth to organize, engage in intellectual

Figure 7.2 Novyi Urengoy youth leaving for motocross behind the workshop of the municipal club "Nord".

(Photo: F. Stammler, 2019)

games and role-play games, invent their own programmes, develop business ideas or just hang out. The municipality also finances the building and basic staff, while the activities are often directed by volunteers. Ideas seen at the centre were born out of practice, some being public service publications such as "How to walk your dog in a clean way in town" or "How to live a healthy lifestyle", one other a project titled "Veterans live next door", in which the participants went to record the life histories of war veterans. In addition, Novy Urengoy has two state-of-the-art sports centres that are as well-equipped and well-staffed as one may find in Moscow, New York or Paris. Co-financed by Gazprom and the municipality, any sports training for youth is free there, and possibilities to develop one's sports ambitions are plentiful. The most talented and ambitious youth may end up as professionals either in the Gazprom-owned sports clubs or elsewhere, although the orientation of the centre is towards promoting sport among the population at large (Reimer 2019). Thus, local youth have more to choose from in comparison to many other cities, yet this has not changed the spirit of transience that the city inherited from its Soviet past.

Kirovsk—the North is for living

Our field site in Murmansk Region was the phosphate mining city of Kirovsk, which is on the list of Russian single-industry towns "with the most difficult socio-economic situation". The population decreased from 43,526 to 26,206 between 1989 and 2019 (Rosstat 2020). In its programme, the municipality groups youth policy together with education, culture and sports. In the most recent programme, the desired social and economic results include the improvement of social services as a whole in Kirovsk, which should translate into an increased level of satisfaction among the residents with municipal social services. Another stated aim is improving conditions for the self-realization and adaptation of young people in society. However, of the total budget of the programme, some RUB 968,000,000 (EUR 11,000,000), only RUB 674,000 (EUR 7500) was allocated directly for the implementation of youth policy, in particular youth initiatives and projects for the first three years. During meetings in Kirovsk, we found that young people do not participate much in the implementation of urban projects financed through the programme. The lack of youth centres, places of leisure, support for young entrepreneurship and attractive employment other than with the principal mining company may make young people think quite seriously of moving to bigger cities. On the other hand, youth in Kirovsk are active in areas not connected with any municipal or company programmes. There are several youth centres or associations established by young people themselves, examples being the alternative cinema club "Art powder", a number of self-organized groups for extreme winter sports and survival training in the forest, as well as a recycling initiative, some of which are described in more detail in Bolotova's chapter (Bolotova, this volume).

Another particular feature of Kirovsk is geographical: of all the cities in our research, Kirovsk is the one which is best connected to Moscow and St. Petersburg, and another big city, Murmansk, is only 200 km away, a three-hour drive. Moreover, several other industrial cities are within an hour's drive, such as Monchegorsk, Olenegorsk, Polyarnye Zory and Kandalaksha. Perhaps most importantly, Kirovsk has a younger and bigger twin city, Apatity, just half an hour's drive away; Apatity has twice the population of Kirovsk and more youth-related activities as well. Apatity also has a youth centre funded by the municipality and that employs seventeen people. A municipal programme called "Involving youth in social practice" aims at improving the socio-economic and organizational conditions for the successful integration of youth into society. The focus is on involving youth more in social activities, promoting career guidance and career aspirations of youth and providing assistance for developing their creative potential. Among the measures contributing to this last goal are bonuses and financial incentives for especially talented children.

Some of the more active youth commute between Kirovsk and Apatity. There is a volunteer association which has 311 registered young people as members. A lot of the volunteers' activities take place in both Kirovsk and Apatity, with these including volunteer "cleaning-up Saturdays" (*subbotniki*), a video festival of extreme sports and youth cinema. During fieldwork in 2019 we learned that some of these activities are co-funded by the mining company, while others are crowd-funded. On balance, the impression from talking to youth active in the cities is that there are many more opportunities where young people can test themselves and realize themselves than merely looking at municipal programmes might suggest.

The principal enterprise of Kirovsk and Apatity is Apatit, part of the PhosAgro company, one of the world's largest producers of phosphate for mineral fertilizers. The company is of crucial importance to both cities. The social orientation of the company has a strong focus on its young employees. For example, the company's youth council takes part in improving the life of young people and has 350 members. In addition to supporting its own employees, the company has financed or co-financed the repair of infrastructure of importance to young people, such as the Kirovsk swimming pool, the palace of culture, sports centres and the downhill skiing resort "BigWood".

Kirovsk is a focal point in the regional identity-making programme, which features the slogan *"na severe—zhit"* ("the North is for living"). The young governor of the Murmansk Region is said to have coined the slogan himself (Novosti Murmanska 2020), effectively focusing attention on people's wellbeing as a principal goal of regional policies. The Kirovsk ski resort and the snow art attraction *snezhnaya derevnya* ("snow village") have adopted the slogan to promote the city as a tourist destination . In the season 2019/2020, the slogan appeared on souvenirs such as t-shirts, on ice art and on posters at the ski resort (Figure 7.3).

On the one hand, this advertising could be seen as a cheap PR campaign to direct attention away from real problems, as suggested in the analysis of the oppositional newspaper *Novaya Gazeta* (Britskaia 2019). On the other hand, during fieldwork in the spring 2020 we noticed among young people in

Figure 7.3 "The North is for living"—slogan for northern wellbeing by the Murmansk Governor, adopted in the ice art of the Kirovsk snow village.

(Photo: F. Stammler, 2019)

Kirovsk a certain sense of pride in and identification with this slogan. Indeed, it is by far not only tourists from outside the region who buy the t-shirts with the slogan; locals wear them too. In an article on place identity in mining towns in Mexico, Harner (2001, p. 660) underlines that "place identity arises when the shared beliefs about place meaning for the majority match the ideological beliefs of those in power". Young people in Kirovsk endorsing the governor's slogan about a place identity attached to the North may be such an example.

The involvement of industry in educating qualified youth for industrial cities

As we have shown in the beginning, the Russian Arctic needs qualified personnel for furthering development. In this sphere, big enterprises working in the region play an important role in educating students for future work in the single-industry towns. In contemporary Russia, that is, after the end of the Soviet Union, positive effects have been noted in those cities where corporate involvement in education is practiced as part of companies' social policies and corporate social responsibility (Romanenko 2019 see also Adams et al., this volume). Companies can play an important role in better tailoring education to the needs of the labour market, by both materially equipping educational institutions and by arranging practice-oriented training for youth by company employees. In our three case cities, the personnel managers of all the companies have mentioned the advantages of educating local people, as they do not require additional adaptation to the harsh Arctic climate and working conditions. We argue that the extent to which company–public partnerships in educating youth are successful has an important influence on the attractiveness of Arctic single-industry towns as places to live for young people. Several examples from our three case sites show how companies engage in education; we witnessed this engagement during field visits to the respective companies and educational institutions.

Kirovsk and Murmansk Regions have paid attention to the seamless integration of the education offered by schools, colleges and industry. The main company in town, PhosAgro Apatit, sponsors a specialized mining orientation already in secondary school to raise interest in local children for the mining sector (for more details see Bolotova, this volume). Murmansk Arctic State University (MASU) now includes a Kola branch office in Apatity, previously part of Petrozavodsk State University. Moreover, the Khibiny Technical College in Kirovsk, previously affiliated with the National Mineral Resources University, has also become part of MASU. In the aforementioned college, PhosAgro Apatit participates both intellectually and materially in the education process: the mining classes are equipped with simulators and technical stations that are almost identical to those used in the company. Students learn the entire process of extraction and enrichment of Apatit phosphate. Some of the teachers are employees of PhosAgro Apatit and give students the benefit of their own work experience. Most of the mandatory work practice undertaken by the students takes place at the company premises in Kirovsk, and most of those who want to work at PhosAgro after graduation find employment there. At a focus group discussion with students of the college, female pupils especially valued the fact that there are now more subjects to choose from than mining. Since 2018, the college has offered hospitality as a profession in a four-year education programme. During a visit to the college the students suggested demonstrating to us in a role play how they learn to work with tourists in the hospitality business: they set up a hotel reception area, arranged a coffee service and provided advice on excursions and help with other questions typically asked by tourists. Bolotova (this volume) shows the importance of such diversification in the education sector.

This degree of partnership is absent in Novy Urengoy. Gazprom established its own technical educational institution, called Gazprom Tekhnikum Novy Urengoy. However, the option of becoming affiliated with a university was ruled out, as the entire YNAO does not have a single university on its territory, and earlier branch offices of larger universities were closed due to poor quality. Gazprom Tekhnikum is part of the personnel division of Gazprom Moscow and educates specialists in seven areas, with a total of 1000 students being trained at the institution. Economics was discontinued as a major subject, so that today all of the education is gas-oriented. It starts already in the tenth grade and continues for another three years. Once a student enters the Gazprom system, the company can follow his or her professional path anywhere in the country and employment options open up throughout all of Gazprom's divisions. Around 90 per cent of the graduates of the Tekhnikum later become Gazprom employees. The two principal Gazprom divisions in Novy Urengoy, Gazprom dobycha Urengoy and Gazprom dobycha Yamburg, contribute most to the budget of the institution, providing the programmes with the latest technology. Parents pay one-third of the tuition fee, Gazprom the rest. Much as in Kirovsk, students can learn the entire process of gas extraction in practice-oriented classes and laboratories, and they spend approximately one year on Gazprom's production site as interns, during which they receive a salary (fieldwork conversation with Director Sergei Yalov, Gazprom Tekhnikum, 23 May 2019).

During our focus group meeting, students presented a balanced view of the city as a place for young people. On the "plus" side were statements such as "in Gazprom there are a lot of different directions for work. That's cool. And we have no trouble getting a job where we spent our internships." "The city develops here well; there is a lot to do here, fun." "It's cool to participate in mind sports (*intellektual'nye igry*), and there are many on offer. And we have a lot of sports facilities here, which is great." On the other hand, the "minuses" sounded like this: "the winter, the darkness, the mosquitoes, the high rate of chronic diseases and heart diseases here". "We need more cultural events, arts, creative arts, youth entertainment. In winter we all sit at home. We should have covered spaces where we can meet." "The new shopping centre is not the best place to hang out: there are drug addicts there, and violence." On the other hand, in a focus group discussion, secondary school graduates highlighted that the city is attractive after school only for those with an interest in the gas industry (focus group discussions in the Tekhnikum, 23 May 2019, and in a secondary school, 22 May 2019).

In Neryungri, Yakutugol' has a centre where they train students in almost 200 professions that are in demand at industrial enterprises. For many years the company has also accepted students from the South Yakutian College of Technology as well as from the Training Centre of the city of Neryungri as interns. Many of them later become employees of the company. Thus, the company helps implement what is called a dual training model, where students get training in college as well as in the company. Both parties try to bring training programmes in the colleges as close as possible to real production conditions. In some professions, students get as much as 50 to 70 per cent of their training at the enterprise, with the acquisition of practical skills directly in workplaces through mentoring.

Since 2004, Kolmar, another coal mining company, has managed to compete with the main coal mining company Yakutugol' and to offer a viable alternative as an employer. During fieldwork, both companies competed for the best young specialists. Kolmar had put up posters in the city advertising themselves as an employer, even offering a financial reward for young people who recommend their friends as potential young specialists. Companies competing for the best youth in town is a good sign of city development. Clearly, demand for qualified labour in Neryungri exceeds supply. Companies try to position themselves as attractive employers and offer a variety of benefits in addition to municipal support. For example, Kolmar is investing in a new suburb for the company's workers that aims to host 2100 young families, with "young" being "under the age of 25" (Meeting with company chief of staff department, 30 July 2018). A specially built school, kindergarten, playgrounds, and healthcare centre show their focus on young workers' social wellbeing (MK Yakutia 2020). Both coal companies, Yakutugol' and Kolmar, are also involved in higher education in Neryungri's technical institute. They contribute to training engineers, help equip laboratories and try to make contacts with the best students through work practice even before their graduation in order to recruit them as employees later.

In the focus group meeting with young staff at Kolmar, the young chief of staff explained why they compete for the best young locals: "For the company, young people are a good deal because they are cheap, and you can teach them what is needed." As in Novy Urengoy, the young people themselves had mixed feelings about life in the city. One statement by a very young mother summarized it philosophically: "We always think it's great where we are not. You just have to want it, then you can do many things here." Her colleague left Neryungri immediately after school and came back recently: "Until you finish school it's great here. But 10 years ago it was much worse than now; there is lots to do if you have money." Her colleague specifies what that means:

> I want to become a mother here; it's safe and everything here is available close-by: schools, kindergartens, social services. And we have a long

holiday: every two years our company pays for our holiday travel in full. Also, children can go to camps at company expense; in 2018, 90 children over the age of fourteen were financed. And, our big plus is nature; that is great in summer. But summer lasts only two months. There is energy in this vastness. During winter, it's running home and to work.

This shows that in our case cities there is strong involvement of industry in education and companies engage extensively in measures, making the place attractive for young people to pursue a career in the fields that the companies represent. Fieldwork conversations with young people from within the industry-dominated system reveal the positive sides of this company engagement. The other side of the coin is that apart from industry there is not much else in the way of a career path for a young person in our case cities—the focus of Bolotova's chapter in this volume.

Discussion

Analysing the implementation of youth policy in mining cities of the Russian Arctic may lead to the conclusion that at the end of the day it is money that matters: the city of Novy Urengoy has the richest industrial company in town and invests a billion rubles in a youth programme, and it is the only town in our comparison with an increase in population. However, it is not yet possible to evaluate the effectiveness of such investments in the long term. Especially in Novy Urengoy, the youngest of the Russian cities in our study, the psychology of transience still creates an image of the city as a place to live in while one works, but not necessarily a place to put down roots. In our fieldwork, it became clear that even in the gas industry, students may plan to "work for a couple years in town, and then leave to the south" (student at Gazprom Tekhnikum, 21 May 2019). This shows that establishing a feeling of a place being "a small homeland" is a difficult task for the future.

In Kirovsk, with its longer history, more young people perceive their city as their permanent home: they may or may not choose to leave it later in life, but not because the place offers a work-only environment. Especially with the city and the company investing in the downhill skiing resort, Kirovsk is slowly starting to develop into a tourist destination, which can become an important factor for economic diversification. Already now this has made the place attractive for entrepreneurs offering low-priced hotel and hostel accommodation, and the ski resort figures prominently in the Russian social media as a hype destination for free-riding (off-piste downhill skiing). A diversified economy may also mean a wider range of options for young people in the community. Establishing hospitality as a field of post-secondary education in Kirovsk in 2018 might contribute to diversification beyond mining.

This observation compares interestingly with observations made by Komu and Adams (this volume) in their case study of the Finnish town of Kolari, where tourism brings more diversity to the town and is a source of employment, while mining is what Haikola and Anshelm (2017) have called a "horizon of expectation". However, in Kolari much of the work in tourism is seasonal and done by workers from the south, who leave once the downhill skiing season is over. Komu and Adams show that such employment is not exactly a dream career for young people. Nonetheless, tourism has the potential to attract youth, because of successful locally run tourism businesses. In this respect, Kirovsk presents a very different case, because the principal investor in tourism is the mining company (see Bolotova, this volume), which also sponsors tourism education. In Kolari, the mining company has made no investment at all in tourism. Rather, Komu (2019) shows that tourism benefits from the infrastructure that was built earlier for the mining industry. Nonetheless, in current discussions about mining and tourism as prospective work, many in Kolari perceive mining as a threat to tourism rather than an opportunity, while in Kirovsk it is the mining company that is the driving force behind the tourism. Thus, the role of the company is clearly more important in Russia than in Finland. We speculate that this may be due to the prominent role of the state in the Finnish system, whereby it provides much more for its citizens' basic and extended needs, their wellbeing and their welfare than in Russia. This being the case, in Russia there is greater need and more space for corporate responsibility. On the other hand, Panapanaan et al. (2003, p. 143) have argued that in Finland companies have revived CSR, and education is a key element in their efforts.

The literature (Dushkova and Krasovskaya 2018 and Bolotova, this volume) and first fieldwork experiences suggest that, in addition to big corporate actors, the young civil society in Kirovsk, one independent of the company and the municipal administration, is the most developed of our three case sites.

Across all cities, young adults achieving economic independence from parents and relatives is an important basic condition of wellbeing. For this to happen, the connection between getting a good education, a decently paid profession and opportunities for subsequent employment is crucial, as is access to comfortable housing for young families.

Due to budgetary constraints and the way the Russian tax system is built, the municipalities often have little room for manoeuvre due to insufficient financial resources. Therefore, various forms of public–private co-financing, such as those outlined above, are crucial for creating the conditions that make single-industry towns attractive to youth for planning their future there. In the same vein, the principal industrial companies in the cities operate assistance programmes for young employees to qualify for bank loans or mortgages.

One could argue that company involvement in youth issues compensates at least partially for the lack of state funding. However, unlike in the Soviet Union, companies withdraw their social involvement and argue that their main social task is paying taxes (fieldwork conversation, Yakutugol', Neryungri 1 August 2018). This is similar in international practice, as for example our fieldwork in Tysfjord (Norway) has shown. There, Heidelberg Cement argues that the Norwegian state takes care of social issues sufficiently, with the company doing its part in the form of taxes and, occasionally, in-kind contributions (fieldwork Ivanova and Stammler, 21 May 2016, interview with Norcem plant manager). The difference is, however, that Western welfare states provide social services to their populations on a high level, whereas in the Russian context in many cases the services only satisfy basic needs, and industries are the only institutions that can augment these. The level of such needs is formulated in Finland, for example, by the Youth Act. Among other things, we have shown in this chapter that the lack of such a cohesive legal instrument leads to stark differences in services, programmes and infrastructure for youth in different Russian Arctic regions.

However, our fieldwork shows that young people will not be satisfied with mere financial benefits. They want to have not only interesting and well-paid jobs, but also a range of other affordances: comfortable living in extreme climatic conditions, suitable housing; a decent standard of education, healthcare, communications and the Internet; functioning transport infrastructure; cultural facilities; and opportunities for realizing their life plans, having a rewarding social life and raising their children. Based on our analysis of diverse legal acts and programmes which are not connected under a general umbrella, we suggest that better coordination of legislation concerning the Arctic and youth could facilitate development towards providing the opportunities enumerated above. Correspondingly, legal norms governing how to attract youth to and encourage them to stay in the Russian Arctic, which are today scattered in diverse regulatory legal acts, should become an integral part of a unified state youth strategy.

In an effort to this end, the youth chambers established under the parliaments of the Arctic regions of the Russian Federation have set out to convince lawmakers to create conditions for the widest possible inclusion of young people in the development of Arctic education, design, innovation and production. This is most urgent in the case of projects related to the infrastructural development of the North and the Arctic.

Conclusion

This chapter on youth in the Russian Arctic has outlined the interface between Russia's federal policy and legislation in the sphere of Arctic youth issues and the local outcome of their operation in three case regions. With the strategic importance of the Arctic for Russia's economy, the starting point was the demand for labour in the Russian Arctic extractive industries, juxtaposed with the tendency of young people to migrate from the single-industry towns studied. Current Russian federal law in the case of both Arctic issues and youth lacks coordination and is fragmented across numerous legislative documents. As such, it does not tackle the problem of youth outmigration from Arctic cities and the demand for qualified youth there in a satisfactory way. Our fieldwork in all three cities with students and educational institutions has shown that training youth to meet such demand is now, to a considerable extent, in the hands of corporate actors instead of being regulated by state legislation. In the absence of federal laws on the Arctic and youth, the Arctic regions can make use of their legislative competences and implement their own policies. We have shown through fieldwork evidence that people and decision-makers in the Russian Arctic have more agency than the image of a strong centralised federal state with a strong president may suggest. Moreover, implementation of current federal youth law in the regions is sometimes fragmentary because of a lack of funding. Some of the laws on the books can therefore be seen as little more than declarations of noble aims.

Our analyses have shown a focus of youth legislation on qualifying personnel for working in extractive industries. Considering UN Sustainable Development Goal 11 on cities as an inclusive, safe. resilient and sustainable space for life, this is too narrow a view. If one wants the Arctic to be a home instead of a place to work, more has to be done to increasing the wellbeing of youth beyond the workplace.

We have analysed how youth policy actors on the city level have approached this topic. In Neryungri, the competition for the best young workers in the two coal mining companies creates an atmosphere of development in town that has the potential to pull along both youth and the municipality. Civil society initiatives and citizen participation in shaping their hometown complement that atmosphere. However, young people still lack choices for their life trajectories in town other than working for industry. In Novy Urengoy the situation is similar, but more so than in Neryungri youth grow up with the feeling of their hometown as being a place to leave, even after starting a career in the gas industry. The municipal programme on developing civil society with money from gas industry taxes aims to overcome that transience, but it may well be that money cannot buy sense of place. The example of Kirovsk has shown that more youth initiative was observable with much less money, but a longer history (also see Bolotova, this volume). Additionally, the promising local identity slogan launched by regional authorities, "the North is for living", might have been embraced by young people to the extent that it strengthens their wish to build a future in Kirovsk.

In all of our case sites we have identified promising initiatives to strengthen civil society and collective agency among young people, agency that creates conditions allowing them to shape their hometowns into the kind of sustainable city communities that they would like to live in. Young people have expressed views on what should change in their cities for the better, and they have opportunities to influence this with their own grassroots initiatives, as our analysis and other chapters in this book show. Their involvement in legal initiatives through youth parliaments can contribute to these developments being backed up by state policies, possibly leading towards better, consolidated youth legislation. However, in the current conditions, marked by rather fragmented state involvement in the Arctic of Russia, it is crucial that industry lives up to its social responsibility and contributes financially not only to educating its future employees, but also to making the cities comfortable spaces in which to thrive.

References

Adams, R.-M., Allemann, L. and Tynkkynen, V.-P. (this volume) 'Youth well-being in "atomic towns": The cases of Polyarnye Zori and Pyhäjoki', in Stammler, F. and Toivanen, R. (eds) *Young people, wellbeing and placemaking in the Arctic.* London: Routledge, pp. 222–240.

Administratsiia goroda Novyi Urengoi (2017) *O vnesenii izmeneniia v postanovlenie Administratsii goroda Novyi Urengoi ot 01.11.2016 № 366.* Available at: http://www. newurengoy.ru/docs/11806-o-vnesenii-izmeneniya-v-postanovlenie-administracii-goroda-novyy-urengoy-ot-01112013-366.html (Accessed: September 26 2019).

Administratsiia Neriungrinskogo Raiona (2016) *Ob utverzhdenii munitsipal'noi programmy «Realizatsiia munitsipal'noi molodezhnoi politiki v Neriungrinskom raione na 2017–2021 gody.* Available at: http://www.neruadmin.ru/elib/aktOMS/normprav/2018/1225_21.08.2018N.pdf (Accessed: September 26 2019).

Administratsiia Neriungrinskogo Raiona (2018a) *O vnesenii izmenenii v munitsipal'nuiu programmu «Obespechenie zhil'em molodykh semei Neriungrinskogo raiona na 2017–2021 goda.* Available at: http://www.neruadmin.ru/elib/aktOMS/normprav/2018/1620_07.11.2018N.pdf (Accessed: September 27 2019).

Administratsiia Neriungrinskogo Raiona (2018b) *O vnesenii izmenenii v prilozhenie k postanovleniiu Neriungrinskoi administratsii ot 14.12.2016 № 1805 « Ob utverzhdenii Poriadka okazaniia dopolnitel'nykh mer sotsial'noi podderzhki obuchaiushchimsia i studentam iz maloobespechennykh semei.* Available at: http://www.neruadmin.ru/elib/aktOMS/normprav/2018/91_25.01.2018n.pdf (Accessed: September 27 2019).

Administratsiia Neriungrinskogo Raiona (2018c) *Ob utverzhdenii munitsipal'noi programmy «Realizatsiia otdel'nykh napravlenii sotsial'noi politiki v Neriungrinskom raione na 2017–2021 gody.* Available at: http://www.neruadmin.ru/elib/aktOMS/normprav/2018/1235_24.08.2018%20N.pdf (Accessed: September 28 2019).

Assotsiatsiia Poliarnikov, R. (2018) *Rezoliutsiia-2018.* St Petersburg: Forumarctic, 120pp. Available at: http://www.forumarctic.com/upload/conf2018/resolution/arctic_2018.pdf (Accessed: July 23 2019).

Assotsiatsiia Poliarnikov, R. (2019) 'Youth Day for the International Arctic Forum', in *Arktika, Nastoiashchee i Budushchee. VII International Arctic Forum.* St Petersburg: forumarctica, 5pp. Available at: https://forumarctica.rcfiles.rcmedia.ru/upload/uf/2d c/2dce026b211c2711d9859a7b31934038.pdf (Accessed: December 16 2020).

Bolotova, A. (this volume) 'Leaving or staying? Youth agency and the livability of industrial towns in the Russian Arctic', in Stammler, F. and Toivanen, R. (eds) *Young people, wellbeing and placemaking in the Arctic.* London: Routledge, pp. 53–76.

Britskaia, T. (2019) '«Na Severe — zhut'» Gosudarstvo osvaivaet Arktiku iskliuchitel'no s voennymi tseliami. Uroven' zhizni v Zapoliar'e stanovitsia vse nizhe', *Novaya Gazeta.* 113th edn, 7 October. Available at: https://novayagazeta.ru/articles/2019/10/07/82266-na-severe-zhut (Accessed: December 10 2020).

Dolgova, A., Ruzakova, V. and Siluanova, L. (2017) *Arktika 18-24-35. Vzgliad Molodykh.* working group report. Moscow: Tsentr Strategicheskikh Razrabotok, p. 52. Available at: https://strategy.csr.ru/user/pages/researches/Arktika_strategy.pdf (Accessed: December 16 2020).

Dushkova, D. and Krasovskaya, T. (2018) 'Post-Soviet single-industry cities in northern Russia: movement towards sustainable development: A case study of Kirovsk', *Belgeo*, (4), pp. 1–24. doi: 10.4000/belgeo.27427.

Federative Council, R. F. (2009) *O gosudarstvennykh merakh po privlecheniiu i zakrepleniiu molodezhi vo vnov' osvaivaemykh raionakh Severa i Arktiki*. Round table discussion outcome. Moscow: Izd. Soveta Federatsii, p. 3. Available at: http://council. gov.ru/media/files/41d44f243f233576911e.pdf (Accessed: December 16 2020).

Fondahl, G., Espiritu, A. and Ivanova, A. (2020) 'Russia's Arctic Regions and Policies', in Coates, K. and Holroyd, C. (eds) *The Palgrave Handbook of Arctic Policy and Politics*. London & New York: Palgrave Macmillan, pp. 195–216. doi: 10.1007/978-3-030-20557-7.

Fondahl, G. et al. (2019) 'Niches of agency: Managing state-region relations through law in Russia', *Space and Polity*, *23*(1), pp. 49–66. doi: 10.1080/13562576.2019.1594752.

Government of the Russian Federation (2014) *Osnovy gosudarstvennoi molodezhnoi politiki Rossiiskoi Federatsii na period do 2025 goda*. Available at: http://government.ru/docs/15965/ (Accessed: April 25 2020).

Government of the Russian Federation (2016) *podprogramma 'Obespechenie zhil'em molodykh semei' federal'noi tselevoi programmy 'Zhilishche' na 2015–2020 gody, Postanovlenie Pravitel'stva RF*. Available at: https://rosreestr.gov.ru/upload/Doc/17-upr/%D0%A2%D0%B5%D0%BA%D1%81%D1%82%20%D0%B4%D0%BE%D0%BA%D1%83%D0%BC%D0%B5%D0%BD%D1%82%D0%B0.pdf.

Haikola, S. and Anshelm, J. (2017) 'The making of mining expectations: mining romanticism and historical memory in a neoliberal political landscape', *Social & Cultural Geography*, pp. 1–30. doi: 10.1080/14649365.2017.1291987.

Harner, J. (2001) 'Place Identity and Copper Mining in Sonora, Mexico', *Annals of the Association of American Geographers*, *91*(4), pp. 660–680. doi: 10.1111/0004-5608.00264.

Heleniak, T. (2009) 'Growth poles and ghost towns in the Russian Far North', in Wilson Rowe, E. (ed) *Russia and the North*. Ottawa: Univ of Ottawa Press, pp. 129–163.

Hill, F. and Gaddy, C. G. (2003) *The Siberian curse: How communist planners left Russia out in the cold*. Washington, DC: Brookings Institution Press.

Komu, T. (2019) 'Dreams of treasures and dreams of wilderness – engaging with the beyond-the-rational in extractive industries in northern Fennoscandia', *The Polar Journal*, *9*(1), pp. 113–132. doi: 10.1080/2154896X.2019.1618556.

Komu, T. and Adams, R.-M. (this volume) Not wanting to be "stuck": Exploring the role of mobility for young people's wellbeing in northern Finland', in Stammler, F. and Toivanen, R. (eds) *Young people, wellbeing and placemaking in the Arctic*. London: Routledge, pp. 32–52.

Konsul'tantPlius, L. information service (2017) *Situatsiia: Severnaia nadbavka inostrantsu, Konsul'tantPlius*. Available at: https://iak.ru/page.xhtml?u=16A53E80 DC574D75B4C3891FD0FFD025 (Accessed: December 10 2020).

Mikhailovskaia, M. (2018a) 'Kak zainteresovat' molodezh' ostavat'sia v Arktike', *Parlamentskaia gazeta*, 6 December. Available at: https://www.pnp.ru/social/kak-zainteresovat-molodezh-ostavatsya-v-arktike.html (Accessed: July 20 2019).

Mikhailovskaia, M. (2018b) 'Molodezhnyi parlament nauchit pisat' zakony', *Parlamentskaia Gazeta*, 6 June, p. online. Available at: https://www.pnp.ru/social/kak-zainteresovat-molodezh-ostavatsya-v-arktike.html

Novosti Murmanska (2020) 'Andrei Chibis rasskazal, kak poiavilsia slogan «Na Severe – zhit'!»', *Murmanskiy Vestnik*, 30 May, p. online. Available at: https://www.mvestnik.ru/newslent/andrej-chibis-rasskazal-kak-poyavilsya-slogan-na-severe-zhit/

O molodezhnoi politike v Iamalo-Nenetskom avtonomnom okruge (2009). Available at: http://docs.cntd.ru/document/895219346 (Accessed: July 29 2019).

Ob okhrane sem'i, materinstva, ottsovstva i detstva (2008). Available at: http://docs. cntd.ru/document/819089973 (Accessed: August 4 2019).

Panapanaan, V. M. et al. (2003) 'Roadmapping corporate social responsibility in Finnish companies', *Journal of Business Ethics*, *44*(2), pp. 133–148. doi: 10.1023/ A:1023391530903.

Parente, G., Shiklomanov, N. and Streletskiy, D. (2012) 'Living in the new North: Migration to and from Russian Arctic cities', *Focus on Geography*, *55*, pp. 77–89. doi: 10.1111/j.1949-8535.2012.00048.x.

Pavlov, S. S. (2018) 'Interview with the director of the Neryungri Institute of Technology'.

Press-sluzhba Il Tumena (2019) *Narodnye deputaty Iakutii utverdili Polozhenie o Molodezhnom parlamente pri Il Tumene, Ministry of Youth Affairs and Social Communications of the Republic of Sakha (Yakutia)*. Available at: https://minmol. sakha.gov.ru/news/front/view/id/2980060 (Accessed: December 10 2020).

Proshchalgin, A. (2018) 'Interview with the evangelical youth pastor'. Available at: Neryungri.

Reimer, O. (2019) 'Interview with the Chief of Youth Administration of Novy Urengoy'. Available at: Novy Urenogy.

Romanenko, K. (2019) *Predpriiatie vse reshaet: kak ustroeno obrazovanie v monogoro-dakh, theoryandpractice.ru*. Available at: https://theoryandpractice.ru/posts/17240-predpriyatie-vse-reshaet-kak-ustroeno-obrazovanie-v-monogorodakh (Accessed: December 16 2020).

Rosstat (2020) 'Chislennost' naseleniia Rossiiskoi Federatsii po Munitsipal'nym Obrazovaniiam'. *Statistics Russia*. Available at: https://web.archive.org/ web/20200822004543/https://rosstat.gov.ru/storage/mediabank/CcG8qBhP/mun_ obr2020.rar (Accessed: December 16 2020).

Russian Federation, S. D. (2019) *Ob obshchikh printsipakh organizatsii mestnogo samoupravleniia v Rossiiskoi Federatsii*. Available at: http://www.consultant.ru/ document/cons_doc_LAW_44571 (Accessed: September 24 2019).

Russian Federation, S. D. (2020) *O gosudarstvennoi podderzhke predprinimatel'skoi deiatel'nosti v Arkticheskoi zone Rossiiskoi Federatsii*. Available at: https://rg.ru/ 2020/07/16/193-fz-ob-arkticheskoy-zone-dok.html (Accessed: December 31 2020).

Severpost, M. (2017) *V Murmanskoi oblasti reshili vossozdat' molodezhnyi parlament, severpost.ru*. Available at: https://severpost.ru/read/55634/ (Accessed: December 10 2020).

Simakova, A., Pitukhina, M. and Ivanova, A. (this volume) Motives for migrating among youth in Russian Arctic industrial towns', in Stammler, F. and Toivanen, R. (eds) *Young people, wellbeing and placemaking in the Arctic*. London: Routledge, pp. 17-31.

Stammler, F. (2010) 'The city became the homeland for its inhabitants, but nobody is planning to die here – Anthropological reflections on human communities in the Northern city', in Stammler, F. and Eilmsteiner-Saxinger, G. (eds) *Biography, shift-labour and socialisation in a northern industrial city – the far north: Particularities of labour and human socialisation : proceedings of the international conference in Novy Urengoy, Russia, 4th-6th December 2008*. Rovaniemi: University of Lapland, Arctic Centre, pp. 33–43. Available at: http://lauda.ulapland.fi/handle/10024/59445 (Accessed: September 8 2017).

Stammler, F. and Ivanova, A. (2016) 'Resources, rights and communities: Extractive mega-projects and local people in the Russian Arctic', *Europe-Asia Studies*, *68*(7), pp. 1220–1244. doi: 10.1080/09668136.2016.1222605.

State archive Murmansk (2012) *Iz istorii razvitiia zakonodatel'stva o «severnykh l'gotakh»*. archival resources summaries. Murmansk: Gosudarstvennyi archiv Murmanskoi Oblasti. Available at: https://www.murmanarchiv.ru/index.php/news/34-publications/327--l-r (Accessed: January 3 2021).

Truntseva, L. (2018) 'Oplata truda. Razbiraemsia v nadbavkakh', *Gudok.ru Russian Railroad newspaper*. 102nd edn, 21 June, p. 1 http://gudok.ru/zdr/171/?ID=1423506 (Accessed: December 10 2020).

Trutnev, I. and Kozlov, A. (2020) *Press-konferentsiia Iuriia Trutneva i Aleksandra Kozlova na temu «Razvitie Arkticheskoi zony Rossiiskoi Federatsii: novye preferentsii dlia biznesa»*. Available at: https://minvr.gov.ru/press-center/news/26952/ (Accessed: January 1 2021).

Il Tumen (2018) *O privlechenii molodykh spetsialistov na gosudarstvennye unitarnye predpriiatiia i v gosudarstvennye uchrezhdeniia Respubliki Sakha (Iakutiia), khoziaistvennye obshchestva*. Available at: http://docs.cntd.ru/document/550103122 (Accessed: April 4 2019).

Volgin, N. A., Shirokova, L. N. and Mosina, L. L. (2018) 'Aktual'nye voprosy razvitiia rossiiskogo Severa: kompensatsionnye i stimuliruiushchie sistemy, napravlennye na privlechenie i zakreplenie naseleniia v severnykh i arkticheskikhb regionakh', *Uroven' zhizni naseleniia regionov Rossii, 208*(2), p. 46.

Vorobiev, A. S. (2019) 'Interview with the coordinator of the youth centre "Nord"'. Available at: Novy Urengoy.

MK Yakutia (2020) 'V Neriungri nachalos' stroitel'stvo novogo kvartala', *MK Yakutia*, 10 September, p. online. Available at: https://yakutia.mk.ru/politics/2020/09/10/v-g-neryungri-nachalos-stroitelstvo-novogo-kvartala.html

Zakonodatel'noe Sobranie YNAO (2016) *Molodezhnyi parlament pri Zakonodatel'nom Sobranii Iamalo-Nenetskogo avtonomnogo okruga, Legislative Assembly of the Yamalo-Nenets Autonomous Okrug*. Available at: https://zs.yanao.ru/about/collegiate/176/ (Accessed: December 10 2020).

Zamiatina, N. Y. (2020) *Gorodskaia Arktika: prostranstva v snegu i doma na merzlote, Go Arctic*. Available at: http://goarctic.ru/society/gorodskaya-arktika-prostranstva-v-snegu-i-doma-na-merzlote/ (Accessed: December 11 2020).

8 The quest for independent living in Finland

Youth shelter as a critical moment in young adults' life courses

Miia Lähde and Jenni Mölkänen

Introduction

In Finland, as in most countries of the Global North, turning 18 signposts juridical adulthood which is accompanied by changes in citizen rights and responsibilities. It also marks a new kind of adult status, one associated with cultural expectations regarding education, commitment to work life, intimate relationships, and independent living.[1] During the first decades of the 21st century, youth policy efforts aimed to prevent social exclusion and promote the participation of young people have contributed to the measures developed from a societal perspective emphasizing rapid transitions and activation, which some young people perceive as oppressive control and therefore unsuited to their actual situations (e.g. Brunila and Lundahl 2020; Mertanen 2020), Alongside these emphases, the perspectives and needs of young people as they forge their independent lives in the phase of 'emerging adulthood' (Arnett 2000; Juvonen 2015, p. 194), and the particularities of this life stage as framed by current political–economic structures and societal transformations (e.g. Miles 2000, pp. 35–48; Aapola and Ketokivi 2005; MacDonald and Shildrick 2018), have remained largely unrecognized in public debate and services.

This chapter addresses the challenges young people face in pursuing independent living, and their views on using the support offered to achieve this independence. It focuses on young people in Finland who have experienced difficulties in their transition to adulthood. We analyze the significance of a third sector service provider, the Finnish Red Cross Youth Shelters, in supporting young people in their quest to achieving an independent adult life. In theoretical terms, the chapter provides a discussion on 'critical moments' (discussed later) by analyzing how affiliation to a particular service or form of support can be critical to young people's wellbeing and further their possibilities of independent living. In practical terms, it contemplates the necessity of services sensitive to the diverse needs prevalent during the emerging adulthood phase.

The conceptualizations, timing and institutional support of youth independence vary historically and culturally. Finnish youth move away from their childhood homes at an earlier age than youth in most other European

DOI: 10.4324/9781003110019-12

countries (Eurostat 2019). Becoming independent, also viewed in Finnish society as a culturally defined ideal, is one of the main reasons for moving out from one's childhood home (Myllyniemi 2017, p. 22; Myllyniemi and Suurpää 2009). Nordic societies and their rearing cultures value early independence, which is reflected in the support provided by the state to enable young people to live separately from their parents after a certain age.[2] The Finnish state supports the process of young people's independence by providing welfare services, including economic subsidies such as study allowances, housing support and unemployment insurance (Myllyniemi and Kiilakoski 2018, pp. 16–17).

Youth services in Finland are based on the Nordic welfare model, emphasizing wide and equal access to the services. They include, for instance, educational, social work, employment, primary and specialized health care services as well as multi-professional counselling and support services (Gissler et al. 2018). Third sector, consisting of non-governmental and non-profit organizations that work alongside the public and private sectors, complements the system of welfare service production (Hiilamo and Saari 2010; Pirkkalainen et al. 2018). It has been argued that despite the state support and a relatively well-covered service system, young people who leave their childhood home early are especially prone to facing economic risks (Berngruber 2016, pp. 194–195; Myllyniemi 2017, pp. 22–23, p. 60). Moreover, inflexible and bureaucratic services, growing dependency on the system and the lack of genuine social encounters have driven exclusion instead of improving youths' inclusion into the welfare regime (Närhi et al. 2013).

Our study is based on a research collaboration between the ALL-YOUTH Research Project and a third sector service provider, the Finnish Red Cross Youth Shelters[3] (Honkatukia et al. 2020). The Youth Shelters operate in the Helsinki metropolitan area (namely the cities of Helsinki, Espoo and Vantaa) and two other big cities in southern Finland (Tampere and Turku). They augment the municipal youth services, which are versatile but also in high demand in these areas compared to many other parts of the country. The shelters are intended to support young people (up to 25 years of age) during times of crisis, such as when they face problems related to family, other social relationships, school and housing, or have unresolved issues about independent living. This aid is designed to be temporary and encompasses a broad range of voluntary and free of charge services, including short-term accommodation, personal and family meetings (Punainen Risti 2017).[4] One of these services under the Kotipolku Project helps young adults aged between 18 and 24 years in the capital metropolitan area to find a home and support their independent living (Kotipolku 2020).[5] Henceforth, we refer to these services by the abbreviation 'YS' (Youth Shelters).

In this chapter, we examine, based on qualitative life course interviews, young people's reasons for seeking support from the YS as well as the kinds of support they found helpful. Our analysis is inspired by a biographical perspective and the concept of a 'critical moment' as a way to compare narratives of transition and capture the importance of life events in trajectories of young people (Thomson et al. 2002, 2004; Thomson 2007; Thomson and Holland 2015) (see the section titled 'Conceptual framework'). By focusing on young persons' affiliation to YS as a 'critical moment' we want to: (1) draw attention to the formation of young people's support needs and the significance of their service experiences in biographical context; and (2) consider how these experiences can inform respective practice with young people. Rather than viewing young people only as vulnerable and marginalized (Brunila et al. 2019, p. 113), such an approach helps one understand them as agents who actively negotiate and manage their lives and identities, and in so doing, participate in and contribute to society in a variety of ways (Honkatukia et al. 2020, pp. 25–28; Thomson et al. 2004, p. 221).

First, we lay out the conceptual framework based on a biographical perspective of the life course transitions of these youth. Then, we present our data and methods, focusing on young people's affiliations to the third sector service YS. Using these 'critical moments' in these young adults' life stories, we explore experiences and events that have produced twists and turns in their transitions and heightened their need for support, and we address how these services can provide the needed assistance. To begin with, we discuss three common underlying themes that manifested in the young adults' narratives, namely insecurities in close relationships, illness or struggle with psycho-social wellbeing and moving residence, which influenced their contact with the YS, and second, we consider the young people's need for support and their experiences of receiving it. We conclude by discussing our results and their possible implications to future studies and the development of services promoting youth independence and wellbeing in Finland.

Conceptual framework

The independency of Finnish youth is typically perceived as increasing emotional, social and economic independence from parents or the familial context of growing up in the transition to independent living. The normative expectation of adultness is primarily marked by autonomy from one's childhood family (Hoikkala 1993, pp. 82–85). This process, which lies at the heart of the youth period (Arnett 2001, p. 134; Miles 2000), is guided by a standardized path, and deviations from it are easily considered as a matter of concern (Furlong 2013). In this chapter, we approach the process(es) of independence as a series of personally negotiated stages and movements over time (Jones 2009, pp. 140–141; Montgomery 2007, p. 283), taking place through interrelated institutionalized transitions associated with the cultural ideals of adulthood and the expected timings of the life course.

There is a consensus that transitions to adulthood have typically become longer and more complex (Furlong and Cartmel 2007), and youths' biographies are increasingly affected by both institutional structuring and freedom of choice. It is, therefore, recognized that the shapes and timings of transition processes vary considerably by social and materially based inequalities (Thomson and Holland 2015, p. 724). For example, young adults leaving care

are somewhat 'forced' to transition to independent living quickly or at once, with no option to return (Kulmala and Fomina, this volume). The ongoing transformations in the global economy and labour market also mean that the notion of achieving adulthood, which is the objective of the transitions, has become ambiguous (e.g. Blatterer 2007; Thomson and Holland 2015; Woodman and Wyn 2015; Cameron et al. 2018). It is argued that the associated instability and insecurity are likely to be the most sharply felt by young disadvantaged people (MacDonald and Shildrick 2013, 2018).

Traditionally, a stable position in working life is highly valued in Finland (Brunila et al. 2013; Aapola-Kari and Wrede-Jäntti 2017), and education is considered a requirement for entering the working life (Aaltonen et al. 2018, p. 5). While research and policy interests have mainly focused on transitions in the context of education and work, research on youth in the 21st century has increasingly touched upon how these transitions intersect with other domains in young people's lives over time, creating the overall form and character of their transitions to adulthood (MacDonald and Shildrick 2018). Developing the ideas laid down in theories of late modernity (e.g. Beck 1992; Giddens 1991), such studies have aimed to document and understand how young people may be experiencing and negotiating themselves and their lives in new social conditions (cf. Du Bois-Reymond 1998).

Our approach to processes of independence is inspired by the biographical perspective and the idea of 'critical moments' developed by the researchers in the longitudinal 'Inventing Adulthoods' study (Henderson et al. 2007). Instead of devising a narrow individual focus or concentrating on policy-related categories, it provides a way to appreciate young people's lives holistically and of seeing also "the limited character of their 'choices' and the critical role of timing in their biographies" (Thomson et al. 2004, p. 237). It emphasizes young people's subjective understandings, feelings and competencies. Adapting this perspective, we join the critique of the normative model, expecting sovereign individuals who are able, in every situation, to evaluate, choose and act (Juvonen 2015, p. 164), and we place at the centre of inquiry individuals who have different, socially and materially based resources and skills with which they orient towards the future (Skeggs 2014, p. 11, pp. 57–59).

According to Thomson (2007, pp. 103–104), the biographical approach encourages one to pay attention to interrelated domains of an individual's life as well as young people's agency and responses to their life events and circumstances. The perspective draws upon the life course theory which identifies four central factors that together have explanatory value in determining unfolding life chances: historical time and place, the timing of lives, linked or interdependent lives and human agency (Elder Jr. 1994; Thomson and Holland 2015, p. 724). It recognizes, as Woodman and Leccardi (2015, p. 711) note, that engagement in one area of life, demands 'an investment of time and energy that is likely to put pressure on the time and energy that is available to invest in other biographical fields at play in a young person's particular social location'. A 'critical moment' can be understood as events or experiences in a person's life which are consequential to their biography; that is, they are defined by their roles in providing biographical momentum and structure (Thomson et al. 2002).[6] The power of the critical moments can be felt at the time, but their significance is often recognized only in retrospect in the narration of the life course (MacDonald and Shildrick 2013, p. 155).

As part of the ALL-YOUTH Project, our study focuses on young adults' constructions of their pasts in one-off interviews and the manner in which these constructions intertwined with their sense of their current lives and related future prospects (Kulmala and Fomina, this volume; Wilson et al. 2007). We take a critical realist stand to the relationship between the 'life as lived' and 'life as told' (Thomson 2007, p. 77), as we are interested in the social circumstances related to life events and young people's experiences of them, even while accepting that knowledge (young people's accounts of themselves as well as our interpretations) is defined by the interactional setting in which they are produced (e.g. the interview schedule and interview style) and, as such, a discursively bound and changing social construction (Lewis-Beck et al. 2004).

Our use of the notion of 'critical moment' is twofold. On the one hand, we utilize it as a tool in data gathering, asking the interviewees to identify events or experiences they themselves found particularly meaningful for their respective life course. On the other hand, our analysis focuses on the affiliation of this research's participants to the YS as 'critical moments', to consider their biographical salience to these young people's transitions to independent living.

Data and methods

We draw upon 17 thematic life course interviews with 18-to-24-year-old young adults in two major urban areas in Finland. The participants were recruited through the YS in collaboration with its staff. They shared one aspect; they all had faced challenges in transition to independence and sought help from the YS. Otherwise, they were a heterogeneous group in terms of life situations, gender (12 females, 3 males and 2 transgenders),[7] ethnicity, family background and social class (ranging from seemingly middle class to socio-economically disadvantaged). Many talked about difficult family circumstances, and some had been in foster care. Some had moved to Finland as children or young persons, and two had a refugee background (Honkatukia et al. 2020).[8] Some had grown up in their current place of residence; however, as is typical for youth transitions in Finland, many had moved towards the urban centres of southern Finland after leaving their childhood homes. Four researchers, including the authors of this chapter, conducted the interviews between March and June 2019. All the interviews were conducted in a place chosen by the interviewee, most often at the YS and sometimes in a café. Pseudonyms are used for the young people, and none of the places, with the exception of the capital (Helsinki) metropolitan area, are named throughout the reporting.[9]

The collaboration with the YS allowed us to interview 'difficult-to-reach' young adults who often decline to participate in such studies. Those who voluntarily participated in the study spoke vividly about their experiences, providing qualitative material rich in descriptions of their social relationships, their educational, work and housing paths and institutional encounters. While the data are not representative of either the youth population in Finland or those who seek assistance from the YS,[10] they provide a multifaceted view of the obstacles to independent living faced by young people in Finnish society and their experiences as users of welfare services. It is also possible that the recruitment process may have led to a positive bias in our data, emphasizing young persons who are predominantly satisfied with YS or confident enough to verbally communicate their life stories.

To encourage the young people to share their stories and reflections of their life courses, we combined a voluntary task-based activity, called the 'lifeline', and thematic life story interviews. As our approach was strongly informed by research ethics and sensitivity to the issues arising in the context of the YS,[11] we assumed that a visual tool could help to facilitate the discussion and allow the interviewees to signal what kinds of things they were ready to expose and talk about with the interviewers (cf. Wilson et al. 2007). We asked each interviewee to draw a lifeline (or 'any visual representation') and jot down on a sheet of paper events or experiences (both positive and negative) that they found particularly meaningful in their life histories. These lifelines provided some structure to the interview conversations, as we invited the participants to elaborate upon their notes of particular moments and then adapted our interview themes to their accounts. The interview schedule included themes of social relationships (family and friends), school, work (other institutional encounters), leisure time, future orientation and Finnish society. We aimed to keep the interviews flexible, allowing the interviewee to control the pace and direction of the conversation. One major advantage of the lifeline was that our interviewees used it to raise sensitive issues (such as illness, substance abuse and their relationships as clients of the social and health services) without us having to ask about them (ibid.).

In the ALL-YOUTH Study, we wanted to take account of young people's feelings in the formation of citizenship experiences and paid attention to the meanings young people gave to their situations and experiences, and to the relationships between different fields of life. We analyzed the interview transcripts by referring to central research themes[12] and timing of life events in terms of 'critical moments' (Honkatukia et al. 2020, p. 11, pp. 21–23)[13]. Our analytic focus in this chapter rests on the interviewees' affiliations with the YS and their consequences on their biographies. Using this information, we identified thematic parallels between these young adults' life situations and need for support, and we considered the role of voluntary non-governmental services in supporting their wellbeing during their transitions to adulthood.

While only 4 interviewees out of 17 marked the YS on their lifelines (Anni, 22 years; Pinja, 20 years; Linnea, 20 years and Nikki, 19 years), most of the interviewees emphasized its positive impacts on their respective situations in their accounts. We observed a slight gender difference in the discourse, in that males placed less emphasis on the importance of mental support and the overall significance of the YS in their lives. All the interviewees were never-theless willing to discuss the services and support they received, which justi-fies including all the interviews in this analysis.

Accounts of entering YS

The reasons and avenues for becoming involved with the YS were varied, such as needing support to maintain a daily rhythm (to go to school, for instance), solving the practical issues associated with navigating the service system, managing to live alone, having a safe space to sleep or discussing one's life situation. The interviewees' contact with the Kotipolku Project in the capital metropolitan area was typically related to resolving housing issues, such as finding one's own home or waiting to move into it and requesting support for independent living in one's own home. Some interviewees had found the service by themselves, while others came upon it with assistance from their family members or a professional in another service (social worker, counselling psychologist or youth worker, for instance).

Looking at the interviewees' lives 'backwards' (starting with their presence at the shelter) revealed three key biographical themes that commonly inter-sected these young people's life stories, informing us about their previous struggles and need for support in pursuit of independent adulthood: fragile social relationships, illness/issues with mental health and moving residence. These themes illustrate how young people's paths to adulthood and their ori-entations to the future are constrained by biographical factors which impact their investments and sense of competence and direct their access to material, social and emotional resources that can help them to build independent adult lives (Thomson et al. 2004; Thomson 2007). They also provide insights on how young people relate to the events and experiences in their immediate relationships and social environments. Even though these themes do not include all the biographical factors associated with the struggle for indepen-dence in our research material, many of the events that interviewees them-selves identified as 'critical moments' in their lifelines intertwined with one or more of these themes.

Fragile family and intimate relationships

Most of the life stories were marked by fragile family relationships and lack of care, security and support. Unpredictable events and discontinuities in social relations, such as illness or death of a parent or friend, were common and had resulted in unforeseeable consequences for young people's transi-tions (MacDonald and Shildrick 2013). Due to the insecurities in their social lives, sometimes persistent since childhood, many interviewees reported trust building issues. Moreover, difficult family situations caused many of them to leave their childhood homes early in comparison to their peers.

We observed different dynamics in social relations, ranging from loud arguments and conflicts to remaining quiet and feeling of emotional distance or of being left alone. Emilia (aged 20) used to quarrel with her mother. She narrated:

> It was a Friday evening in 2015. I was 16. We had a really miserable relationship with my mother. We did not get along and she threw me out of the house that Friday evening. I had a mobile phone, and I drove with my motorbike to the woods and thought there what to do. Finally, I found this place [YS] on the web, and finally, I stayed there.

Emilia remembered that relations between herself and her mother had become strained after she started seventh grade. She was the oldest of the siblings and felt that her mother was always reprimanding her. According to Emilia, it was actually her mother who had told her to move out during their arguments. Her father lived abroad, and she did not know her grandparents. Furthermore, their family had lived abroad because of her father's work and she had not built enduring relationships in Finland. At the time of crisis, she did not know where to go or who to turn to. 'There were no safe adults or anyone like that in my life at that time', she related. Emilia's observations and experiences reveal that in order to feel secure, it is necessary to know people one could rely on and ask for support. Moving from one place or country to another can leave a young person without the much-needed social network required for such a transition.

Sara (21 years) also reported having arguments with her mother.

> I was 18 at the time. My boyfriend at the time also lived at our place. We had a somewhat unbalanced situation at home because my mother had just taken her ex back. I did not get along with my mother's ex and (...) [f]inally, the ex, who had not even moved back to our place, kicked my boyfriend out of the house (...) We fought all the time with my mother and I was also concerned about my sisters. Mum was working 24/7. I went to the web and clicked on a button and got a call from here [YS]. I came to the crises section and as my boyfriend had been thrown out, I had decided that I will move out also.

Sara's account highlighted her tensions with her mother and her mother's spouse. Sara felt that her mother was working all the time, and she was worried about her younger sisters' wellbeing and physical integrity.

These accounts contradict the normative life course narrative, which typically includes the presence of a childhood family, usually with one's biological parents, and thereby securing one's material, social and emotional being (cf. Jones 2009, p. 163). Instead of the satisfactory support required during their growing-up years, the young adults highlighted instability in family relations due to separation, death and problems with money or intoxicants, for instance. Sometimes, lack of parenting resulted from a parent falling short of resources because of his/her own problems or difficulties with other family members. Nikki (19 years) related how she had not learnt social skills at home.

> Well, I think that my past is a very relevant matter in relation to my independence. See, I come from a bit of a bad family. There were problems with alcohol, I was not allowed to go out and nobody could come to our place ... as if I was not allowed to be visible and heard. I did not learn those normal skills required for independence, let alone social skills, at all.

These feelings of 'invisibility', loneliness and not getting recognized were also present in the other interviewees' accounts. For example, after the sudden death of her father, Veera (aged 20) felt that she had mainly become a babysitter for her younger siblings in a reconstituted family.

In addition to family relations, peers and romantic relationships played a significant role in young adults' lives. Many had struggled for years with self-esteem and trust issues because of the bullying they had experienced at school. Valtteri (24 years) contacted the YS because he had become homeless after breaking up with his short-term girlfriend with whom he had shared an apartment. He did not maintain contact with his mother, and his father had died when he was 6 years old. His grandmother was dead too. He himself had a little daughter whom he had not seen. When Valtteri and his mother moved to a new town, he did not stay there long. He left alone to return to the place he had moved from because his friends lived there. Eventually, he was enrolled in child care services at the age of 16. Like Emilia, Valtteri did not have close relationships with his childhood family, and all his brothers and sisters had been moved to foster homes.

Fragile family relationships and tensions or the experience of being ignored at home sometimes resulted in accelerated steps toward independence and leaving of the childhood home, highlighting the need to find meaning and support from other social relationships. In most cases, the young adults' reflection on their frail relations, wellbeing and resources with regard to initiating independent living were related and emphasized their need for support. A few interview accounts nevertheless depicted family relations as unproblematic and unrelated with support needs such as requiring help with life management or acquiring an apartment. Moreover, despite some history of family problems, a few interviewees narrated that their parents had supported their process of independence by helping with living arrangements, for instance. Some had come to the YS with a parent or due to the parent's advice.

Illness and psycho-social wellbeing issues

While health concerns as such were not a reason for these young people to contact the YS, they overlapped intricately with their life stories and current challenges. In addition to their own health problems, many had encountered illness and bereavement in their immediate social relationships, which were often conveyed in the interviews in a mundane manner although they struck us, the interviewers, as out of the ordinary and distressing (MacDonald and Shildrick 2013). The most common type of ill-health issue recounted by the youngsters themselves was depression, which was often associated with anxiety or other mental health problems.

Viivi (24 years) had entered the Kotipolku Project via access to outreach youth work[14] in the midst of her move from one house to another; she was not able to find a new home as quickly as she had hoped. She had been simultaneously seeking a place for a work experiment or rehabilitation. Viivi related her difficulties with her health issues which were also intertwined with her social relations.

> I had been diagnosed with Asperger later in my life, and I feel that it is the reason why I have had all kinds of problems. When I was 16, I looked for psychiatric help for the first time. I think there were multiple issues, such as little sisters and a new living place after moving to a new town. After moving I realized that I was unwell in some way. I was stuck. I had no means of going forward and I did not know what was wrong.

Anni (22 years) told us about her rheumatism.

> I went to study to become a waitress although its [rheumatism's] symptoms were apparent. I never had to stay away from school because of rheumatism although we had almost one year of work practice. Of course, there were periods during my leisure time when I was more sick and more tired than usual and I had occasional pains.

Her rheumatism had required attention throughout her life. She had tried many different medications, because some of them had resulted in side-effects. In addition to rheumatism, she was diagnosed with depression in secondary school.

> I had anxiety and depression, and I used to cut myself. Well, my mother noticed and contacted the school nurse who guided me to a city youth psychiatric centre. I went a lot to therapy; however, I was able to go to school. My workload at school was reduced by 20% and I was allowed to come to the school later so that I could sleep enough. I used to stay up and cry, so of course, I was tired in the morning.

The manner in which people categorize and experience certain events depends on their social environments (Thomson et al. 2002, p. 342). Viivi's narrative indicates that she remained confused and was unable to solve issues and 'go forward'. She felt that one reason for her problems was the developmental disorder, which affected nonverbal communication and caused difficulties in social interaction since school age. At home, she did not want to burden already tired parents with her own problems; instead, she suffered quietly and retreated into herself.

The youngsters we interviewed puzzled about how they would manage in their work lives because of their struggles with disorders, depression and other mental health problems, illnesses, medication requirements and hospital visits, all of which could disrupt their attention at school. For example, Anni's narrative demonstrated how helping someone with an illness to recover requires the attention of not just the person undergoing that illness, but also his/her family, friends and institutions.

In addition to illnesses and disorders, some of the young adults confessed to problems with substance use. For example, Valtteri had been partaking alcohol and other intoxicants from secondary school onwards. After completing basic education, he had attended school only irregularly because of his alcohol use and the requirement to travel between the town the school was located in and the town in which his friends resided. He used to drink alcohol with his friends and consequently skipped school. Interviewed at the age of 24, he was studying in vocational education and training. He regretted 'missing' so many years of his early adulthood because of a reckless lifestyle resulting from his early freedom that independent living resulted in after his leaving care. 'But then, it can't be changed. Forward, forward', he said.

Aleksi (aged 21), who received his own home via the Kotipolku Project, explained the following:

> Well, I am in a messed up situation. All kinds of things have happened recently. Last summer, I left the army. Then I worked for a while ... I do not know ... I was distressed about everything and I lost interest in life. Well then I started to use drugs. I had to go to the rehabilitation but I did not stay there. It was too short a visit.

Aleksi had quit military service after a short while, disappointing his father, who had told Aleksi that he should not come home and instead live elsewhere. He had then moved in with his sibling, and his mother had helped him to contact the YS Project in order to find a place of his own. Aleksi's drug use seemed to be related to his feeling of not knowing what to do. He was unable to agree to long-term commitments required for the military, a working life and rehabilitation.

Overcoming an illness requires specific attention and care, and, as in the case of substance use, these efforts take time away from school, which is a crucial institutional marker of life transitions. The interview accounts also show how young adults negotiated their everyday life in poor health conditions, which arose from their difficult transition experiences and complicated their social activities and their relations with the welfare service system. On the other hand, substance use appeared to offer some form of sociability, which the young people found more appealing in their life situations, particularly when their institutional duties were stressful or difficult to deal with.

Moving residence

The third unifying theme in the young interviewees' narratives of seeking support at the YS involved physically moving from one town or neighbourhood to another or out of their childhood homes. These young people moved for different reasons, and sometimes, the moving was dictated—even forced—by external factors. Notably, moving to one's own home was configured as a turning point in the process of independence for the interviewees, who reported experiencing mixed feelings and practical challenges. Jerome, who stayed at the YS for some time while waiting to get his own apartment after having moved to town, felt that having a place to call his own had been a starting point for his new life and helped to solve some of his troubles. For Nora, moving into her own home made life easier as she escaped from the violent atmosphere of her childhood home.

> The moving went well. I lived in this town and went to school in another. Sometimes, it took as much as two hours to travel to the school. The moving helped the travel-related stress and I did not have that feeling of responsibility that I had experienced in my [childhood] family. I was allowed to be independent. I only had to take care of the apartment.

Although Nora experienced struggles at the beginning of her independent living phase (we will explore these later in this chapter), she felt that living alone helped her to concentrate on her studies and her life in general.

The young adults reported struggles with the support system, as they were unaware about how to navigate it. These issues resulted in problems with money and paying rent. Some expected that their moving to live alone meant that they would decide every aspect of their lives as they pleased and would live exactly as they wished. However, they found their daily routines in disarray and resorted to binge-watching movies and programmes online, much like the practice in their childhood homes.

Young people who immigrate to Finland alone when under-aged experience mixed feelings of hope and despair as well as opportunities, uncertainties and responsibilities (Honkasalo et al. 2017, p. 7). Moreover, some have had to move several times within Finland and adapt to changing locations and social relationships on each such occasion. Jusuf (21 years), for example, had moved four times during the four years he had stayed in Finland.

> I came to Finland in 2015 and I lived in a town in eastern Finland. Then I moved to a different town to study the Finnish language, but I did not have many friends there. Then my social workers told me that I had too long a way to travel to school and so I moved to a new town in eastern Finland. It was better in that town because it contained people from the same country I had belonged to.

His narrative revealed that in the beginning of his stay in Finland, his life was structured and conditioned by Finnish laws and institutions. In accordance with observations from studies focusing on young asylum seekers, Jusuf's movements from one centre to another or to a school housing were suggested by his social workers (Piitulainen 2016, pp. 17–18), and he felt that he had made more friends after moving to a bigger town, where he was able to play football with them in his free time (Maiche 2017, p. 22). These accounts also suggest that institutions and official networks may not always work in the best interests of the young person or fail to provide adequate support, especially to children or young adults who are trying to plan their future without recourse to trusted intimate relationships.

Aada moved away from her childhood home at the age of 16 to escape an extreme situation; she was fleeing from a personal threat. This was the first time she had moved homes after having lived for some periods in a hospital ward. Once she was alone in her student flat in the new town, she was shocked by the lack of friends and loneliness and became depressed because she had no one to turn to. After relocation, she maintained contact with the outreach youth workers and the adolescent psychiatric clinic that had directed her to the YS. She needed support mainly because of the intimidation which continued despite her changing homes, which made her feel afraid to move out alone. She was sometimes escorted to the YS from her home by car. The start of her independent life included a difficult study year with many absences from school, which was partly exacerbated by sleep problems confusing her daily rhythm.

An exploration of the various reasons for entering the YS revealed that obstacles to independent living, such as homelessness, often result from diverse dynamics and accumulated troubles associated with social relations, health issues and/or encounters with institutional structures, such as the service system. For most of our interviewees, these experiences and dealing with them had created delays and twists and turns in transitions in education and from school to work. They recognized that these issues affected their possibilities to make choices and long-term plans in different areas of life and made them dependent on external support, often against their will, increasing uncertainty about their own abilities and future prospects. For some of the interviewees, moving to the big cities of southern Finland had temporarily added to challenges of managing personal life but also had helped to find a combination of services where one's support needs were met.

Enabling spaces and possibilities for support

The focus on the YS as a critical moment in the youngsters' biographies highlights the kinds of help the young adults needed and found useful in their quest to live independently. It draws attention to the role of professionals and different aspects of the service system as part of young people's life stories and thus contribute to improved understanding of more opportunities for providing support (Thomson 2007, p. 73, 104).

As pointed out earlier, the Finnish support system is diverse and includes public, private and third sector services. The interviewees, who had been involved in several services with different sectors, had accumulated both positive and negative experiences of service encounters. They pointed out aspects that had not worked well for them, such as situations when there was confusion about service provision or when there was a break in provision (Munford and Sanders 2016). In this section, based on the young adults' narratives, we highlight the assistance they found helpful in their life situations and in relation to their future plans.

One of the recognized problems within the Finnish support system involves its fragmentation (Perälä et al. 2012). Viivi had become aware of this fragmentation when she had moved from one place to another.

> The worst experience in mental health care was when I moved [to a different town]. I was on a sick leave at the time. They wrote me an admission note stating that my sick leave should be continued. I came to the new place and started the therapy. My papers had not arrived at that time. Then, my papers arrived and the therapist told me that you do not belong here, go away, we are not a bank, go to work. That was the most negative experience that I have ever had and nobody helped. I was like, well, where do I go?

Viivi remembered this experience as 'the worst one'. She had moved to a new place and was abandoned without support. She had just lost her apartment in the town she previously lived in. Eventually, she lived with her previous roommate for half a year, and in her own words, 'twiddled [her] thumbs and just wondered what I am going to do now. Everything collapsed and I stayed out'.

The fragmentation and overlapping of services over different sectors or municipalities weakens their influence due to problems with information transfer and the provision of the services by only one sector (Pekkarinen and Myllyniemi 2012, pp. 26–27; Perälä et al. 2012). These oversights lead to situations in which only the symptoms are treated, while the actual problems persist. In addition, some cases do not receive any assistance at all, as in Viivi's example. Her words describe how the existing everyday structures and practices 'collapsed', a word often used in situations when people's experiences change (Lawler 2010). Viivi felt excluded, was not able to do much and was forced to stay 'outside'. This kind of support structure excludes certain people, and also maintains inequalities and divisions between those inside and the others outside it (Saari 2019, p. 24). Finally, in geographical terms too, the support system is uneven. For example, when Viivi relocated to the urban capital area, she felt that the doctors there possessed the professional experience to help her.

In general, the young adults described that the YS had supported them and their everyday coping concretely by providing temporary accommodation, helping them to fill the different forms required by the bureaucratic system or advising them on carefully managing their salary, study allowance or living expenses. The young adults also mentioned that they had been able to discuss their life situations, possibilities and future hopes honestly at the YS.

The young people's reflections, however, showed that they lacked concrete knowledge of the workings of the support system, the kinds of support available to them and their rights (see also Kulmala and Fomina, this volume). Nora explained that she was quite lost as to how things in her life should work.

> Clearly, I was not able to handle all that is required when one is independent; money for rent, water and electricity bills, telephone, food and clothes … everything. I wanted someone who had gone through the same things to advise me on what to do.

The importance of concrete and practical support was also highlighted by Emilia.

> I got help from here [YS] with money issues, like with an alimony, and do I have any possibility to apply for some monetary support … I did not know anything about those things …We filled in applications for KELA[15] and copied them together (…) And everybody were really friendly here and I got all the stamps and envelopes.

Researchers have pointed out that the institutional system itself can marginalize young people: understanding the service system requires multiple skills, time management, flexibility and the patience to wait one's turn (Aaltonen et al. 2015; Brunila et al. 2018). In addition, people with difficulties are often quiet and have been silenced (Aaltonen and Kivijärvi 2017, p. 17; Atkinson et al. 2003, pp. 71–72). Indeed, the young adults highlighted situations where they had remained quiet although they were unsure about whether their affairs had been handled well by the experts. Anni explained how her visits to the rheumatism clinic were not always pleasant.

> I did not get what I wanted or what belonged to me at the doctor's. I always felt that we disagreed with the doctor, and because I was a young girl, I always left with the feeling that everything is not well, and I called crying to my mum that I don't know what I will do.

The service system is built on the assumption that individuals who seek assistance are able and willing to talk and describe their situations. For example, Aada reported that she was only able to talk about her issues with the psychiatrist after seeing them over a longer time period: 'When I did not know how to talk, there was no point of going there.' Valtteri stated that, 'One gets help when one asks', emphasizing the importance of asking for the required service and support, and sometimes, even knowing what to ask.

In spite of their negative experiences, the young adults themselves offered ideas for relevant support. A very concrete and practical suggestion came from Nora, who suggested that civic skills should be taught in elementary school: 'They should teach very practical civics in comprehensive school. Like how to fill in KELA forms.'

In addition to the practical help, the YS supported the young adults by their very presence, discussing their lives, thoughts, fears, hopes and plans. As Emilia put it,

> ... and the most important thing was that when I said something, the people [at the YS] did not just agree with me; they were able to argue against me and discuss the issue from different perspectives ... It was important that they listened but did not pet one's head. And when I called here, it did not have to be for some really devastating event and I was still allowed to come here. That was important.

These reflections show that the young people appreciated the concrete and holistic support that the official sector-based service system of the welfare state may not always be able to provide. The young adults appreciated talking, listening and being offered diverse and realistic perspectives. These perspectives were not only related to their work or educational plans, but also concerned their identities, competencies, social relations, the kinds of lives they wanted to live and the possibilities.

Simply being aware that a safe alternative such as the YS existed for young adults to contact at any time improved their feelings of security. Pinja, who had been provided child protection under social services and later moved from the hospital to live by herself, described the following:

> It was important that the support was not tied to the office hours. In addition, it was not always about problem solving but just hanging around, being quiet or chatting about everyday matters. It was important to know that there was a place that one could contact and go, even if one was living by himself/herself.

In addition to knowing that the support was not restricted to office hours, the fact that the YS provided a relaxing atmosphere that allowed socializing and did not necessarily include problem solving was considered to be very relieving. When needed, the YS served as a temporary, safe and drug-free accommodation between 5 p.m. and 10 a.m. daily, and many of the interviewees had made use of these facilities. Unlike the Russian care leavers interviewed by Kulmala and Fomina (this volume), who experienced their targeted services as stigmatizing, the participants in our study did not discuss social stigma in association with the YS, which are open and accessible to anyone.

The narratives of the young adults also inform us that the safe and reliable workers they found at the YS strengthened their ability to cope during difficult life situations, supported their life choices and increased their sense of safety. The employees of the YS were significant adults who could provide the young people with alternative perspectives of difficult situations and life in general, helping them to recognize new aspects of themselves and believe in their dreams and the possibility of building their own lives (see also Kulmala and Fomina, this volume). At the same time and according to the

YS ethos, many young people felt they had been able to re-connect and build a more positive rapport with their parents. The positive accounts of getting help from the YS thus point to the young adults' subjective experiences of being recognized as people with their own personalities, special needs and life situations, and not only in terms of their problems.

Our data also show how young adults became familiar with fragmented services and project-like working lives through their contacts with the support system (cf., e.g., Puuronen 2014a, p. 30). Viivi noted,

> I would like to complain about project work because most of the help that I get at the moment from The Girls House and outreach youth work is paid as projects. I have been talking with these workers, and in both cases, the relationships with them ended because their project funding had ended.

The young adults felt that it was tiring to establish new relationships with new people. As Anni described: 'I am tired of explaining my situation over and over again and getting to know new people as people [in the services and support system] change ... it is just so f****** tiring'.

The young adults' reflections show that they appreciated long-term relations with support personnel (Gissler et al. 2018). It was important for them that one or more adults know about their situations. However, not all the workers and young people got along. Jerome indicated that he told the YS volunteers 'I do not like you' upon recognizing that he did not get along with them. Jerome was able to express his boundaries despite the challenges he had confronted as he had come to Finland alone as a young teenager. Such a comment and the underlying need to draw boundaries informs us about agency that tends to be restricted and controlled in the institutional service context (Foucault 2005; Goffman 1969; Honkatukia et al. 2020, pp. 54–55; Kallio 2019, p. 169). It demonstrates the importance for young people to be allowed to express themselves and to be heard and thus maintain their autonomy in navigating the services and managing themselves.

Conclusions

Researchers in youth studies have called for more nuanced accounts of the manner in which young people's experiences in different spheres of life, and in relation to services, interact (Aapola and Ketokivi 2005; MacDonald and Shildrick 2007; Thomson 2007; Suurpää 2009; Aaltonen and Kivijärvi 2017). In this chapter, we focused on biographical interviews that captured the dynamic processes of independence and experiences of young people on the threshold of adulthood who are typically targets of policy interventions and whose voices are not so well heard in the society (cf. MacDonald and Shildrick 2018). As in the case of the youth population in general, the study participants were a diverse group of people from different social backgrounds, some more committed to society than others. Their struggles of independence are not unique to Finland; they run parallel with the lives of the youths in other Nordic countries (e.g. Ekman 2020, p. 2; Paulsen et al. 2018; Jørgensen et al. 2019). Thus, despite the local data analysis and framework setting, the study contributes to a wider debate and policy emphases on young people's wellbeing and opportunities for meaningful citizen engagement in the context of social sustainability. Our observations also resonate with those of Kulmala and Fomina in this book, addressing, in a different cultural context, how young adult's agency is informed by social circumstances and negotiated at the junction of the past, present and future.

The biographical perspective illustrates how the social conditions of growing up shape young people's wellbeing and how unpredicted life events or insecure social relationships may impact on their chances of building independent lives. At the same time, it helps to identify the characteristics that shape contemporary transitions to adulthood more generally. We focused on the significance of the YS as a 'moment' in each of our interviewees' life courses, which is 'critical' given its consequences on their lives and identities as well as its revelations of the needs topical in the life stage of emerging independence and the kinds of support the young adults found appropriate. Our analysis of the YS as a 'critical moment' also highlights the importance of safe adult relationships and sound emotional support on the way to independent living and points to the possibility of providing an emotional social safety net.

The young interviewees' motivations to seek help at the YS and their support needs varied depending on the specific issues affecting them, such as finding a residence, searching for more long-term support with everyday life management or seeking psycho-social wellbeing (e.g. self-confidence or a sense of security). In addition to family background and other immediate relationships, their challenges were influenced by health issues and the changes associated with moving from one location to another.[16] As these various issues were often intertwined and had been accumulating over the years, their challenging life situations could not be set right with short-term solutions and a narrow approach to defining problems. It seems, nevertheless, that the youths' sense of autonomy and competence benefited from the support and nurturing provided by the YS, which granted them time to resolve their issues and a low threshold for seeking aid. In doing so, the system recognized a person in his/her specific social circumstances.

One of the most prominent features in our data concerned the fragility of social relationships defining the transition to adulthood. Many of the interviewees had suffered from bullying at school or had taken responsibility for themselves, and sometimes even their younger siblings, at an early age. Tellingly, most of them acknowledged the significance of reliable adults and impartial space at the YS, where they could think about their identities and take steps towards independent living. Despite their familiarity with the other welfare services, they noted that the kind of practical and emotional support provided by the YS was essential to their wellbeing. As per many interviewees' life narratives, their affiliations to the YS signified positive change and personal growth.

According to our findings, health concerns and health-related behaviours, which have not been studied as key strands influencing youth transitions (MacDonald and Shildrick 2013), require particular attention as part of the current day processes of independence in terms of how the personal struggle and social disadvantage related to health problems interact with social structures and the uncertainties in youths' lives (Brunila et al. 2020; Brunila and Lundahl 2020). Mental health problems, which are sometimes associated with other illnesses or addiction, were prominent in our data. They required active attention from the young people, influencing both their investments and sense of agency in different spheres of life, and reducing their possibilities for smooth transitions from education to work and independent living. Depression, for example, which is increasingly diagnosed among the youth (Filatova et al. 2019), both reflects and adds to anxieties about the work life and is therefore related to social expectations and pressures targeting young people.

According to our data, the subjective sense of managing by oneself lies at the core of achieving independence for young people (Honkatukia et al. 2020). However, becoming independent is not a transition towards being self-sufficient; instead, it involves the reshaping of the social relationships and mutual dependencies between people (Cameron et al. 2018). Our findings show that the YS workers had an important role in helping the young people to strengthen social ties in their lives. Moreover, the young adults' appreciation of the holistic encounters at the YS suggests that to support wellbeing and independent coping, a third sector service based on voluntary engagement may be able to provide a sense of personal recognition and timely support in a way that the fragmented, often burdened government services cannot necessarily offer. Importantly, these service encounters had allowed the young people to regard themselves as socially included and competent adults who have resources and capabilities for coping (see also Thomson et al. 2004; Juvonen 2015; Munford and Sanders 2016; Gissler et al. 2018).

Altogether, our study highlights the need for more flexible and holistic forms of services for Finnish youth transitioning to independent living compared to the simple focus on quick moves through education to working life. Improved biographical understanding based on the views and experiences of the young adults themselves and their accounts of the diverse paths towards adulthood could provide insights into developing practices to support young adults in their quest for independence.

Acknowledgements

We are especially thankful to all the young people who took part in our study as well as the staff members of Red Cross Youth Shelters who helped us with the recruitment of the study participants. We are also grateful to the editors of the book, other reviewers of the manuscript and Päivi Honkatukia for their insights. We thank Stefan Kirchner for his help and expertise in German social law. The research was funded by the Strategic Research Council at the Academy of Finland under the ALL-YOUTH Project (decision no. 312689).

Notes

1 A crucial change for a small proportion of youth involves the end of the statutory right to alternative care. This change permits after-care with a personal service plan, the age limit of which was recently changed from 21 to 25 years by the amendment to the Finnish Child Welfare Act (Sosiaali- ja terveysministeriö, 2020).

2 In Finland, child benefit ends when a person turns 17, while the German model, for example, allows parents, under certain terms, to continue applying for the child benefit until the child turns 25 (Bundeskindergeldgesetz 2009).

3 https://www.redcross.fi/help-and-support/support-emergency-youth-shelter

4 In 2018, the Youth Shelters supported 1,054 young adults and 1,791 parents. They served 5,370 individual, parent, or network negotiations (Nuorten turvatalon vuosi 2018: 5).

5 The Kotipolku Project was launched in 2018 as a response to an increasingly difficult housing situation for young adults under 25 years in the capital metropolitan area (Kotipolku 2020).

6 Julie Cruikshank (1998) coined the term 'epitomizing event' in the context of oral history studies in anthropology. She also worked with biographical narratives as a research method.

7 We made our own assumptions based on the interviewees' names and self-presentations. We did not ask them about their gender identification preferences unless the issue was brought up by the interviewee.

8 The young adults were contacted by a worker of the YS, who presented the initial interview request. Only persons aged over 18 years were contacted.

9 We are ethically committed to protect the privacy of the interviewees and avoid any harm to them. Thus, locations other than the capital metropolitan area (Helsinki, Espoo and Vantaa) are unspecified, and instead, we refer to 'big city' or larger areas (e.g. 'eastern Finland').

10 Almost two-thirds of the research participants were women even though about half of those who reach out to the YS are male. Despite the staff's efforts to motivate young men to participate, many rejected the opportunity.

11 These concerns were reflected throughout our recruitment and research process. We ensured that the interviewees were aware of their rights, as well as the availability of the YS staff to discuss any aspect of the interview that might upset them. Ethical approval was obtained from the Tampere University Research Ethics Committee. Informed consent was given by the participants after all the steps of the procedure were explained to them.

12 These central research themes were participation in society, relationship with societal institutions, social relations and the importance of (one's own) home.

13 As our definition of critical moments (or the turning points in each account) encompasses both events and experiences emphasized by the young people themselves and what we as researchers identified as critical, it is important to note that these two may differ (Thomson and Holland, 2015: 727).

14 Outreach youth work in Finland takes place under regional state administrative agencies and expands on youth work by reaching youth under 29 years old in their own surroundings. It is a 'detective form' of youth work, which seeks to identify needs for services and supports their delivery with a youth-centred ethos (Puuronen 2014b).

15 KELA refers to the Social Insurance Institution of Finland.

16 There were also other social or socio-structural factors, which influenced some of these young people's challenges of independence such as discrimination based on ethnicity and racialized identity.

References

Aaltonen, S., Berg, P. and Ikäheimo, S. (2015) *Nuoret luukulla-Kolme näkökulmaa syrjäytymiseen ja nuorten asemaan palvelujärjestelmässä*. Nuorisotutkimusseuran/ Nuorisotutkimusverkoston julkaisuja 160. Helsinki: Nuorisotutkimusverkosto/ Nuorisotutkimusseura.

Aaltonen, S. and Kivijärvi, A. (2017) *Nuoret aikuiset hyvinvointipalvelujen käyttäjinä ja kohteina* [Young adults as users and objects of welfare services]. Nuorisotutkimusseuran/Nuorisotutkimusverkoston julkaisuja 198. Helsinki: Nuorisotutkimusverkosto/Nuorisotutkimusseura.

Aaltonen, S., Kivijärvi, A. and Myllylä, M. (2018) Työn ja koulutuksen ulkopuolella olevien nuorten aikuisten koettu hyvinvointi. *Yhteiskuntapolitiikka 84* (3), pp. 301-311. Available at: http://urn.fi/URN:NBN:fi-fe2018110147048.

Aapola, S. and Ketokivi, K. (2005) *Polkuja ja poikkeamia – Aikuisuutta etsimässä* [On and off the beaten tracks – searching for adulthood]. Nuorisotutkimusseuran julkaisuja 56. Helsinki: Nuorisotutkimusseura.

Aapola-Kari, S. and Wrede-Jäntti, M. (2017) 'Perinteisiä toiveita, nykyhetkeen kiinnittyviä pelkoja-nuoret pohtivat tulevaisuutta', in Myllyniemi, S. (eds) *Katse tulevaisuudessa: Nuorisobarometri 2016*. Valtion nuorisoneuvoston julkaisuja, *56*, pp. 159-176.

Arnett, J. J. (2000) 'Emerging adulthood: A theory of development from the late teens through the twenties', *American Psychologist*, *55*(5), pp. 469-480. doi: 10.1037/0003-066X.55.5.469.

Arnett, J. J. (2001) 'Conceptions of the transition to adulthood: Perspectives from adolescence through midlife', *Journal of Adult Development*, *8*(2), pp. 133-142. doi: 10.1023/A:1026450103225.

Atkinson, P., Coffey, A. and Delamont, S. (2003) *Key themes in qualitative research: Continuities and changes*. Walnut Creek: Rowman Altamira.

Beck, U., (1992) ' From industrial society to the risk society: Questions of survival, social structure and ecological enlightenment', *Theory, culture & society*, *9* (1), pp. 97-123.

Berngruber, A. (2016) 'Leaving the parental home as a transition marker to adulthood', in Furlong, A. (eds) *Routledge handbook of youth and young adulthood*. 2nd edn. New York: Routledge, pp. 209-214.

Blatterer, H. (2007/2009) *Coming of age in times of uncertainty*. New York: Berghahn Books.

Brunila, K., Hakala, K., Lahelma, E. and Teittinen, A. (2013) 'Avauksia ammatilliseen koulutukseen ja yhteiskunnallisiin erontekoihin', in Brunila, K., Hakala, K., Lahelma, E. and Teittinen, A. (eds) *Ammatillinen koulutus ja yhteiskunnalliset eronteot*. Gaudeamus Helsinki University Press, pp. 9-16.

Brunila, K., Honkasilta, J., Ikävalko, E., Lanas, M., Masoud, A., Mertanen, K. and Mäkelä, K. (2020) 'The Cultivation of Subjectivity of Young People in Youth Support Systems', in Nehring, D., Madsen, O. J., Cabanas, E., Mills, C. and Kerrigan, D. (eds) *The Routledge International Handbook of global therapeutic cultures*. Abingdon, Oxon: Routledge. doi: 10.4324/9780429024764.

Brunila, K. and Lundahl, L. (2020) 'Introduction', in Brunila, K. and Lundahl, L. (eds) *Youth on the move: Tendencies and tensions in youth policies and practices.* Helsinki: Helsinki University Press, pp. 1–14. doi: 10.33134/HUP-3-1.

Brunila, K., Mertanen, K., Ikävalko, E., Kurki, T., Honkasilta, J., Lanas, M., Leiviskä, A., Masoud, A., Mäkelä, K. and Fernström, P. (2019) 'Nuoret ja tukijärjestelmät haavoittuvuuden eetoksessa', *Kasvatus, 50*(2), pp. 107–119. Available at: http://urn. fi/URN:NBN:fi:ELE-017380263.

Brunila, K., Mertanen, K., Tiainen, K., Kurki, T., Masoud, A., Mäkelä, K. and Ikävalko, E. (2018) 'Essay. Vulnerabilizing young people: Interrupting the ethos of vulnerability, the neoliberal rationality, and the precision education governance', *Journal of the Finnish Anthropological Society, 43*(3), pp. 113–120. doi: 10.30676/ jfas.v43i3.82737.

Bundeskindergeldgesetz (BKGG) [Federal Child Benefit Law] of 28 January 2009, Bundesgesetzblatt (BGBl.) [Federal Gazette] 2009 I 142 and 2009 I 3177, as amended by Article 5 of the Law of 1 December 2020, Bundesgesetzblat 2020 I 2616. Available at: https://www.gesetze-im-internet.de/bkgg_1996/BJNR137800995. html#:~:text=Im%20Fall%20des%20%C2%A7%202,Lebensjahres%20 gew%C3%A4hrt (Accessed: December 8 2020).

Cameron, C., Hollingworth, K., Schoon, I., van Santen, E., Schröer, W., Ristikari, T., Heino, T. and Pekkarinen, E. (2018) 'Care leavers in early adulthood: How do they fare in Britain, Finland and Germany?', *Children and Youth Services Review, 87*, pp. 163–172. doi: 10.1016/j.childyouth.2018.02.031.

Cruikshank, J. (1998) *The social life of stories: Narrative and knowledge in the Yukon Territory.* Lincoln and London: University of Nebraska Press.

Du Bois-Reymond, M. (1998) 'I don't want to commit myself yet': Young people's life concepts', *Journal of Youth Studies, 1*(1), pp. 63–79. doi: 10.1080/13676261. 1998.10592995.

Ekman, T. (2020) *Country sheet on youth policy in Sweden. Partnership between the European Union and the Council of Europe in the field of youth.* Available at: https:// pjp-eu.coe.int/documents/42128013/63134234/Sweden_Country-sheet_April2020. pdf/835bf2e1-674c-e117-590a-55f673e08c72 (Accessed December 15 2020).

Elder, G., Jr. (1994) 'Time, human agency, and social change: Perspectives on the life course', *Social Psychology Quarterly, 57* (1), pp. 4–15. doi: doi:10.2307/2786971.

EUROSTAT, (2019) *When are they ready to leave the nest?* Available at: https:// ec.europa.eu/eurostat/web/products-eurostat-news/-/EDN-20190514-1 (Accessed: May 15 2020).

Filatova, S., Upadhyaya, S., Kronström, K., Suominen, A., Chudal, R., Luntamo, T., Sourander, A. and Gyllenberg, D. (2019) 'Time trends in the incidence of diag-nosed depression among people aged 5–25 years living in Finland 1995–2012', *Nordic Journal of Psychiatry, 73*(8), pp. 475–481. doi:10.1080/08039488.2019.1652342.

Foucault, M. (2005/1975) *Tarkkailla ja rangaista.* Keuruu: Otavan Kirjapaino Oy. Original Michel Foucault (1975) *Surveiller et punir: Naissance de la prison.* Paris: Gallimard.

Furlong, A. (2013) *Youth studies: An introduction.* London: Routledge.

Furlong, A. and Cartmel, F. (2007) *Young people and social change: New perspectives.* 2nd edn. Maidenhead: Open University Press.

Giddens, A. (1991) *Modernity and self-identity: Self and society in the late modern age.* Stanford: Stanford University Press.

Gissler, M., Kekkonen, M. and Känkänen, P. (2018) *Nuoret palveluiden pauloissa: Nuorten elinolot -vuosikirja 2018* [Tangling with the services: Young people's living conditions – yearbook 2018]. Helsinki: Terveyden ja hyvinvoinnin laitos. doi: http://urn.fi/URN:ISBN:978-952-343-200-0.

Goffman, E. (1969) *Minuuden riistäjät. Tutkielma totaalisista laitoksista.* Helsinki: Marraskuun liike. Original Goffman, E. (1961) *Asylums: Essays on the social situation of mental patients and other inmates.* New York: Anchor Books.

Henderson, S., Holland, J., McGrellis, S., Sharpe, S. and Thomson, R. (2007) *Inventing adulthoods. A biographical approach to youth transitions.* London: Sage.

Hiilamo, H. and Saari, J. (2010) *Hyvinvoinnin uusi politiikka: johdatus sosiaalisiin mahdollisuuksiin.* Diakonia-ammattikorkeakoulun julkaisuja. Tampere: Juvenes Print Oy.

Hoikkala, T. (1993) *Katoaako kasvatus, himmeneekö aikuisuus? Aikuistumisen puhe ja kulttuurimallit.* Helsinki: Gaudeamus.

Honkasalo, V., Maiche, K., Onodera, H., Peltola, M. and Suurpää, L. (2017) 'Johdanto', in Honkasalo, V., Maiche, K., Onodera, H., Peltola, M. and Suurpää, L. (eds) *Nuorten turvapaikanhakijoiden elämää vastaanottovaiheessa. Nuorisotutkimusverkosto/Nuorisotutkimusseura Verkkojulkaisuja 2016* (120), pp. 7–8. Available at: http://www.nuorisotutkimusseura.fi/images/julkaisuja/nuorten_turvapaikanhakijoiden_elamaa_vastaanottovaiheessa.pdf (Accessed: August 30 2020).

Honkatukia, P., Kallio, J., Lähde, M. and Mölkänen, J. (2020) *Omana itsenä osa yhteiskuntaa: Itsenäistyvät nuoret aikuiset kansalaisina.* [A part of society in one's own right – Young adults becoming independent as citizens] University of Tampere: Faculty of Social Sciences. Available at: http://urn.fi/URN:ISBN:978-952-03-1731-7 (Accessed: December 9 2020).

Jones, G. (2009) *Youth.* Cambridge: Polity Press.

Jørgensen, C. H., Järvinen, T. and Lundahl, L. (2019) 'A Nordic transition regime? Policies for school-to-work transitions in Sweden, Denmark and Finland', *European Educational Research Journal, 18*(3), pp. 278–297. doi: 10.1177/1474904119830037

Juvonen, T. (2015) *Sosiaalisesti kontrolloitu, hauraasti autonominen. Nuorten toimijuuden rakentuminen etsivässä työssä.* Nuorisotutkimusseuran julkaisuja, 165. Helsinki: Nuorisotutkimusseura.

Kallio, K.P. (2019) 'Elettyä kansalaisuutta jäljittämässä: Kansalaisuuden ulottuvuudet Nuorisobarometrissa', in Pekkarinen, E. and Myllyniemi, S. (eds) *Vaikutusvaltaa Euroopan laidalla. Nuorisobarometri 2018.* Helsinki: Finnish Youth Research Network/Youth Research Society, pp. 167–182. Available at: https://tietoanuorista. fi/wp-content/uploads/2019/03/NB_2018_web.pdf (Accessed: December 10 2020).

Kotipolku (2020) Available at: https://www.punainenristi.fi/kotipolku (Accessed: March 17 2020).

Kulmala, M. and Fomina, A. (this volume) Planning for the future: Future orientation, agency and self-efficacy of young adults leaving care in the Russian Arctic', in Stammler, F. and Toivanen, R. (eds) *Young people, wellbeing and placemaking in the Arctic.* London: Routledge, pp. 196–221.

Lawler, A. (2010) 'Collapse? What collapse? Societal change revisited', *Science, 330*(6006), pp. 907–909. doi: 10.1126/science.330.6006.907

Lewis-Beck, M. S., Bryman, A., and Futing Liao, T. (2004) *The SAGE encyclopedia of social science research methods (Vols. 1–0).* Thousand Oaks, CA: Sage Publications, Inc. doi: 10.4135/9781412950589 (Accessed: December 20 2020).

MacDonald, R. and Shildrick, T. (2007) 'Street corner society: Leisure careers, youth (sub)culture and social exclusion', *Leisure Studies, 26*(3), pp. 339–355. doi: 10.1080/02614360600834826.

MacDonald, R. and Shildrick, T. (2013) 'Youth and wellbeing: experiencing bereavement and ill health in marginalised young people's transitions', *Sociology of Health and Illness, 35* (1), pp. 147–161. doi: 10.1111/j.1467-9566.2012.01488.x.

MacDonald, R. and Shildrick, T. (2018) 'Biography, history and place: Understanding youth transitions in Teesside', in Irwin, S. and Nilsen, A. (eds) *Transitions to adulthood through recession: Youth and inequality in a European comparative perspective.* London, UK: Routledge, pp. 74–96.

Maiche, K. (2017) 'Nuoret turvapaikanhakijat ja luottamuksen rakentuminen arjen käytännöissä', in Honkasalo, V., Maiche, K., Onodera, H., Peltola, M. and Suurpää, L. (eds) *Nuorten turvapaikanhakijoiden elämää vastaanottovaiheessa.* Helsinki: Unigrafia, pp. 19–24.

Mertanen, K. (2020) *Not a single one left behind: Governing the 'youth problem' in youth policies and youth policy implementations.* Helsinki: University of Helsinki. Available at: https://helda.helsinki.fi/handle/10138/320548 (Accessed: December 15 2020).

Miles, S. (2000) *Youth lifestyles in a changing world.* McGraw-Hill Education. Buckingham: Open University Press.

Montgomery, H. (2007) 'Moving', in Kehily, M. J. (ed.) *Understanding youth: Perspectives, identities and practices.* London: Sage, pp. 283–312.

Munford, R. and Sanders, J. (2016) 'Understanding service engagement: Young people's experience of service use', *Journal of Social Work, 16*(3), pp. 283–302. doi: 10.1177/1468017315569676.

Myllyniemi, S. (2017) 'Johdanto & Tilasto-osio', in Myllyniemi, S. (ed.) *Katse tulevaisuudessa. [Looking to the future]. Nuorisobarometri 2016. Nuorisotutkimusseuran/Nuorisotutkimusverkoston julkaisuja 189.* Helsinki: Nuorisotutkimusseura, Nuoriasiain neuvottelukunta, julkaisuja 56 and Opetus- ja kulttuuriministeriö, pp. 5–10.

Myllyniemi, S. and Kiilakoski, T. (2018) 'Tilasto-osio', in Pekkarinen, E. and Myllyniemi, S. (eds) *Opin polut ja pientareet. Nuorisobarometri 2017. Nuorisotutkimusseuran julkaisuja 200.* Helsinki: Nuorisotutkimusseura, Nuorisotutkimusverkosto, Valtion nuorisoneuvosto, julkaisuja 58 and Opetus- ja kulttuuriministeriö, pp. 11–117.

Myllyniemi, S. and Suurpää, L. (2009) '*Nuoret aikuiset itsellisyyden ja riippuvuuden noidankehässä', Hyvinvointikatsaus 1/2009.* Helsinki: Tilastokeskus, pp. 11–14.

Närhi, K., Kokkonen, T. and Mathies, A.-L. (2013) 'Nuorten aikuisten miesten osallisuuden ja toimijuuden reunaehtoja sosiaali- ja työvoimapalveluissa', in Laitinen, M. and Niskala, A. (eds) *Asiakkaat toimijoina sosiaalityössä.* Tampere: Vastapaino. pp. 113–145.

Paulsen, V., Höjer, I. and Melke, A. (2018) 'Editorial 'leaving care in the Nordic countries'', *Nordic Social Work Research, 8*(1), pp. 1–7. doi: 10.1080/2156857X.2018.1520805.

Pekkarinen, E. V. K. and Myllyniemi, S. (2012) *Nuorten elinolot vuosikirja 2012. Lapset ja nuoret instituutioiden kehyksissä.* Helsinki: Nuorisotutkimusverkosto and Terveyden ja hyvinvoinnin laitos.

Perälä, M-L, Halme, N. and Nykänen, S. (2012) *Lasten, nuorten ja perheiden palveluja yhteensovittava johtaminen.* Helsinki: Terveyden ja hyvinvoinnin laitos. Available at: http://urn.fi/URN:ISBN:978-952-245-529-1 (Accessed: April 20 2020).

Piitulainen, M. (2016) 'Kansainvälisen suojelun periaatteista ei voi alaikäisten kohdalla tinkiä', in Honkasalo, V., Maiche, K., Onodera, H., Peltola, M. and Suurpää, L. (eds) *Nuorten turvapaikanhakijoiden elämää vastaanottovaiheessa.*

Nuorisotutkimusseuran and Nuorisotutkimusverkoston julkaisuja *120*, pp. 14–18. Available at: http://www.nuorisotutkimusseura.fi/images/julkaisuja/nuorten_turvapaikanhakijoiden_elamaa_vastaanottovaiheessa.pdf (Accessed: September 20 2020).

Pirkkalainen, P., Mohamed, A. and Aaltio, I. (2018) Third sector hybridization and migrant integration: Cases of two migrant youth organizations in Finland. *Electronic Journal of Business Ethics and Organization Studies, 23*(2). Available at: http://ejbo.jyu.fi/pdf/ejbo_vol23_no2_pages_24-33.pdf (Accessed: January 18 2021).

Puuronen, A. (2014a) 'Millä saa pääsylipun yhteiskuntaan?', in Gretschel, A., Paakkunainen, K., Souto, A.M. and Suurpää, L. (eds) *Nuorten yhteiskuntatakuusta nuorisotakuuseen. Nuorisotakuun arki ja politiikka.* Nuorisotutkimusseura/Nuorisotutkimusverkosto, julkaisuja 150, verkkojulkaisuja 76. Helsinki: Unigrafia, pp. 28–32.

Puuronen, A. (2014b) *Etsivän katse. Etsivä nuorisotyö ammattina ja ammattialan kehittäminen – näkökulmia käytännön työstä* [On the lookout. Outreach youth work as a profession and development of the professional field – aspects of practical work]. Helsinki: Nuorisotutkimusverkosto/Nuorisotutkimusseura. Available at: http://www.nuorisotutkimusseura.fi/images/julkaisuja/etsivan_katse.pdf (Accessed: April 20 2020).

Saari, J. (2019) *Hyvinvointivaltio eriarvoistuneessa yhteiskunnassa. Toimi-hankkeen selvityshenkilönraportti30.1.2019.* Availableat:https://vnk.fi/documents/10616/5698452/Selvityshenkil%C3%B6+Juho+Saaren+raportti+30.1.2019+-+Hyvinvointivaltio+eriarvoistuneessa+yhteiskunnassa (Accessed: March 14 2020).

Skeggs, B. (2014) *Elävä luokka* [Self, Class, Culture]. Tampere: Vastapaino.

Sosiaali- ja terveysministeriö, (2020) *Lastensuojelulain muutokset 1.1.2020.* Available at: https://stm.fi/lastensuojelulain-muutokset (Accessed: March 14 2020).

Suomen Punainen Risti (2017) *Aikalisä riitoihin, yösija tarvitsevalle - Nuorten turvatalo perustettiin 1990.* Available at: https://www.punainenristi.fi/uutiset/20170511/aikalisa-riitoihin-yosija-tarvitsevalle-nuorten-turvatalo-perustettiin-1990 (Accessed: March 14 2020).

Suurpää, L. (2009) *Nuoria koskeva syrjäytymistieto: avauksia tietämisen politiikkaan.* Helsinki: Nuorisotutkimusseura ja Nuorisotutkimusverkosto. Available at: http://www.nuorisotutkimusseura.fi/images/julkaisuja/syrjaytymistieto.pdf (Accessed: January 31 2019).

Thomson, R. (2007) 'A biographical perspective', in Kehily, M.J. (ed.) *Understanding youth: Perspectives, identities and practices.* London: Sage, pp. 73–106.

Thomson, R., Bell, R., Holland, J., Henderson, S., McGrellis, S. and Sharpe, S. (2002) 'Critical moments: Choice, chance and opportunity in young people's narratives of transition', *Sociology, 36*(2), pp. 335–354.

Thomson, R. and Holland, J. (2015) 'Critical moments? The importance of timing in young people's narratives of transition', in Wyn, J. and Cahill, H. (eds) *Handbook of children and youth studies.* Singapore: Springer, pp. 723–733. doi: 10.1007/978-981-4451-15-4_35.

Thomson, R., Holland, J., McGrellis, S., Bell, R., Henderson, S. and Sharpe, S. (2004) 'Inventing adulthoods: A biographical approach to understanding youth citizenship', *The Sociological Review, 52*(2), pp. 218–239. doi: 10.1111/j.1467-954X.2004.00466.x.

Wilson, S., Cunningham-Burley, S., Bancroft, A., Backett-Milburn, K., and Masters, H. (2007) 'Young people, biographical narratives and the life grid: young people's accounts of parental substance use', *Qualitative Research*, 7(1), pp. 135–151 doi: 10.1177/1468794107071427.

Woodman, D. and Leccardi, C. (2015) 'Time and space in youth studies', in Wyn J., Cahill H. (eds.) *Handbook of children and youth studies*. Singapore: Springer, pp: 705–721. doi: 10.1007/978-981-4451-15-4_34.

Woodman, D. and Wyn, J. (2015) 'Class, gender and generation matter: Using the concept of social generation to study inequality and social change', *Journal of Youth Studies 18*(10), pp. 1402–1410. doi: 10.1080/13676261.2015.1048206.

9 Planning for the future

Future orientation, agency and self-efficacy of young adults leaving care in the Russian Arctic

Meri Kulmala and Anna Fomina

Introduction

What goals and expectations do young adults leaving alternative care[1] have for the future in the Russian Arctic? Do they plan? If so, how far does their thinking extend? How do they see their chances to influence their future? What are the major factors in the social context that influence future planning? This chapter explores the expectations of young people who transition from different forms of alternative care into independent adult life (see also Lähde and Mölkänen in this volume).

According to Nurmi (1991, p. 1), thinking and planning for the future are particularly important for young people for several reasons. Firstly, young adults are faced with a number of normative age-specific tasks, most of which concern expected life span development and which require thinking about the future. Secondly, young adults' future-oriented decisions, such as those related to career, lifestyle, and family, have a crucial influence on their later adult life. Thirdly, how young adults see their future plays an important part in their identity formation. Moreover, if young people have experienced challenges and hardships in their life, this also affects how they see their future (also Lähde and Mölkänen, this volume).

As Massey et al. (2008, pp. 424–442; 445–445) write in their review article, a number of studies shows that family context has a great influence on adolescents' future-oriented planning. It does so in terms of parents and children having similar life goals and aspirations, for instance, on education. Maternal support is shown to be related to educational expectations and self-efficacy. Findings concerning the influence of parental socio-economic status or the ethnicity or gender of an adolescent seem to be ambiguous (see Massey et al. 2008, pp. 424–442). One can yet assume that (the lack of a stable) family context has a particular influence on the planning of young adults who have lived in alternative care and whose journey to adulthood, thus, is undertaken against a backdrop of difficult life experiences and sometimes amidst unsupportive family relationships (Hiles et al. 2014, p. 1). Research has indicated that among the young adults who have had such severe adverse experiences as alternative care placement or maltreatment in their lives 'future orientation' and 'planning' promote ability to cope with hardships (Appleton 2019,

DOI: 10.4324/9781003110019-13

p. 4). Thus, imagining one's future with a sense of control over one's life can be considered as a resource its own right: developing such a mastery over one's life provides one with resilience (Hitlin and Kirkpatrick Johnson 2015).

We examined the plans and aspirations of 43 care leavers in two regions of North-West Russia where we, together with young peer-researchers, conducted biographical interviews from December 2018 to October 2019.[2] We analyse how these young adults orientate themselves to the future and their perceptions of how far they control their own future. Our main concept is "subjective agency" involving: (a) perceived capacities; and (b) perceived expectations of what life holds in store. We understand agency as the ability to plan and make related decisions, while a sense of agentic ability refers to young adults' own perception of their ability to master their lives (i.e. self-efficacy). Agency is embedded, on the one hand, in the past; on the other, it is orientated towards the future and the present (Emirbayer and Mische 1998). Building on this conceptual framework, our central question lies in the self-understanding of a young person about her/his ability *at a given moment* to influence *their future*. This present self-perception is obviously informed by *past life experience*, as also Lähde and Mölkänen show in their chapter in this volume.

In addition to our investigation on (future) orientation and agentic ability, we ask what factors the observed orientations interconnect. Through our investigation, we aim to understand the conditions that could facilitate the development of self-efficacy which could contribute further to the resilience of these young adults to cope with various challenges. In our understanding, young adults who make plans exercise their agency and they do this within both enabling and constraining structures and in relationship to other people and in the context of their personal histories (cf. Viuhko 2020, pp. 45–46; also Lähde and Mölkänen, this volume).

In this chapter, we first shortly introduce how support systems for young adults leaving care work in the Russian context. Then we present our theoretical framework and methodology. The empirical analysis that follows is structured by two modes of future orientation found in our study: those who plan and dream ahead and those who show little future orientation or refuse to plan. The analysis is connected to different modes of agency and is followed by a discussion of the external factors that affect these modes of agency.

Russian care leavers and aftercare support

Russia is undergoing massive child welfare reform in line with global trends to dismantle residential care and develop community-based services for families and children and alternative care in foster families. The reform stems from the common understanding that residential care leads to weak social adaptation and social exclusion. One of the priorities of the reform is to develop aftercare support services for young people who transition into their independent life. In Russia, young adults are eligible for such support until the age of 23. A critical moment was in 2014, when the Decree #481 (Decree #481 2014) came into force and transformed the residential children's homes into family support centres that were assigned with new tasks, including preventive and support services for birth and foster families and aftercare services for care leavers. Care in an institutional setting can be provided only as a last resort and on a short and temporary basis. (Kulmala et al. 2021b)

Care reform changed drastically in terms of where children deprived of parental care could be placed. Between 2005 and 2019, the number of residential care institutions decreased by two-thirds (Tarasenko, 2021, p. 102). Instead, children are increasingly placed in foster families: if in 2000 only 1% were placed in foster families, by 2017 the share had risen to 28%. Meanwhile, the share of residential care dropped from 27% to 8% (Biryukova and Makarentseva 2021, p. 32). Additionally, there are a number of children villages in which several foster families often live together (Kulmala et al. 2021b). Thus, the family context—or alternative care placement—might have a significant influence on the future orientation of a young person. In our analysis, we view individual context as a micro-level external factor.

According to Decree #481 each family support centre has to have a specific department working with care leavers. This work ideally includes providing informational, legal, psychological services and personal support. Yet, in reality the centres are often able to provide only minimal support and NGOs provide many kinds of significant supplementary and complementary support (Kulmala et al. 2021d; also Lähde and Mölkänen, this volume, for the Finnish context). NGOs are often forerunners in terms of developing new working practices and approaches as well as services (Kulmala et al. 2021d). For instance, NGOs do valuable work in terms of different aftercare programmes from the so-called 'youth houses' (*dom molodezhi*) to practice independent living to psychological and juridical counselling. NGO-run (and often state-funded) programmes with volunteers act as individual mentors for young people in alternative care and are also currently spread widely throughout the country. There are, however, vast regional differences in Russia in how the public aftercare services function in practice and how developed the non-state provision of services is. Some regions are, for instance, more open to NGOs than others (Skokova et al. 2018; Tarasenko 2021). In our analysis the regional infrastructure we considered to be a meso level external factor.

Care leavers in Russia are supported in many ways by the federal-level social policy, which we view as a macro-level external factor. Generally, the benefits that young people leaving care receive from the state include one-off and monthly payments, as well as subsidies for housing and communal services. In case a care leaver inherited no real estate from their birth parents, (s) he has the right to get a state-sponsored apartment, which, in five years, becomes her/his own property. The apartment is usually located in the municipality where the care leaver officially resides (registration). This is sometimes problematic if a care leaver has built her/his independent adult life in one place, due to studies for instance, but is registered in another place, for instance due to the place of residence of her/his birth. There are some regions where these state-sponsored apartments are all located in certain residential districts, or even in certain buildings, where all the care leavers are then settled (Abramov et al. 2016).

In the sphere of education, care leavers have the right to obtain two secondary vocational degrees and one university degree free-of-charge before they turn 23 years old. While they study, young care leavers receive financial and material support (such as a bursary, money for public transportation and personal hygiene products and clothes) and a place in a dormitory. State policy heavily encourages care leavers to study. Consequently, most of them choose to study but the regional educational infrastructure, societal stigma and lack of individual counselling largely shape this choice. Too often these young adults are "sent" to study in certain less distinguished schools, learning skills related to certain low-paid professions due to a sense that they are unable to do anything better. (Kulmala et al. 2021a). In some regions of Russia there are quotas in certain educational institutions for care leavers (Abramov et al. 2016). After their graduation and up to the age of 23 years, care leavers have the right for six months to receive a targeted unemployment benefit equal to the average salary in the given region, which is higher than ordinary unemployment benefits. At the same time, there is very little support for finding work.

Despite generous state benefits, care leavers face a wide range of problems, many of which are structural in their nature, as discussed in this chapter. One serious obstacle in their life is that children left without parental care carry pretty strong societal stigma in Russian society. As Iarskaia-Smirnova et al. (2021) showed, care leavers are regularly presented in the Russian press as "hopeless criminals" with addictions and unable to adjust to "normal" adult life (see also Khlinovskaya-Rockhill 2010). They are often depicted as "bad learners" and because of that recommended to go "less demanding" schools and professions (Kulmala et al. 2021a) and as viewed as "scroungers" upon the above-described state benefits (Abramov et al. 2016).

Theoretical framework: agency, projectivity, and self-efficacy

In this chapter, we focus on the future orientation and plans of the young adults leaving care to see how they perceive their own ability to make choices and have control over their future. We conceptualize this as subjective agency. What matters to us in our empirical investigation is to what extent and under what conditions young adults exercise their agency when making their future plans and what circumstances enable or restrict this ability. Each individual exercise agency to a certain extent (cf. Hitlin and Elder Jr 2007, p. 185), but some people have more, some less (individual) capabilities or (structural) possibilities to (not to) act or exercise power (Viuhko 2020, p. 44; Hitlin and Kirkpatrick Johnson 2015). Vulnerability somewhat limits agency, but "even those without power have the ability to make decisions though they face severe consequences for those choices" (Hitlin and Elder Jr 2007, p. 185). This ability may vary in different spheres of life.

Even if planning for the future might be an issue of individual decision, as Lähde and Mölkänen also show clearly in this volume, it is structured by state policies and other societal context and individual life histories, which all contribute to a set of repertoires of possible choices. Importantly, planning is structured by social norms: young adults' transition to independent life is heavily directed by normative expectations of a certain kind of path, deviations from which are often seen as somewhat alarming (Furlong 2012; also Lähde and Mölkänen, this volume). Agency is also exercised in relation to other social actors (Viuhko 2020, p. 46): it is a dialogical process by and through which actors immersed in temporal passage engage with others within collectively organized contexts of action (Emirbayer and Mische 1998, pp. 973–974).

In our analysis special attention is paid to care leavers' own perceptions of their agentic ability. Our central concept is self-efficacy: the sense of control over one's life and the ability to see the causal influence of one's own decisions and choices. Developing such a mastery connects with resilience and wellbeing (Hitlin and Kirkpatrick Johnson 2015; Massey et al. 2008, p. 422). With the explicit focus on the future orientation of young adults, relevant to us is what Mustafa Emirbayer and Ann Mische (1998, p. 970) conceptualized as "projectivity", which involved the ability to imagine alternative future trajectories of action, in which received structures of thought and action may be creatively reconfigured in relation to actors. To understand this creative reconstructive dimension of agency, we must focus upon how agentic processes give shape and direction to *future* possibilities (Emirbayer and Mische 1998, p. 984). Yet all forms of agency have a simultaneous internal orientation towards the past, present and future, and thus are temporally embedded in the flow of time. The ways in which people understand their own relationship to the flow of time *make a difference* to their actions (Emirbayer and Mische 1998, p. 973). In this way, we might expect that the past hardships that young adults leaving care have experienced in the process of losing their parents, inform their future orientations and the sense of their own ability to control their lives. Tied together, they might have important life-course consequences.

We build on our previous analysis of the educational choice of young adults leaving care for which we applied (through some modifications) the conceptualization of agency by Hitlin and Elder Jr 2007; see Kulmala et al. 2021a). Two dominant strategies of among young adults in our research to orientate themselves to their future career were found: (1) "long-term planning" connected with "life course agency"; and (2) "not-to-plan" connected with "pragmatic agency" and associated temporally proximate decisions. The life course mode was combined with a sense of control over one's life and the ability to see the causal influence of one's own decisions and choices, while those young adults engaged with the pragmatic mode saw little chance of influencing their educational trajectory. The authors concluded that providing support and care that would promote the development of such an agency is highly important since strong control feelings accumulate in many spheres of life and thereby contribute to overall wellbeing. Yet, as in research by Appleton (2019), Barratt et al. (2019) and Hung and Appleton (2016) on young adults transitioning from alternative care, several young adults in our study expressed their intentions not to plan for the future from the career perspective. In this chapter, we focus more widely than education, turning to expectations and aspirations for the future. We also add one more case study region and re-conceptualize the above-mentioned two strategies as two different orientation categories.

Orientation to the future in accordance with certain expectations, aspirations and goals is what Hitlin and Kirkpatrick Johnson (2015) labelled as the "power of looking ahead". Following the authors, we assume that believing in one's ability to influence one's life is crucial in building a long-term life strategy. Similarly, as for Lähde and Mölkänen in the Finnish context in this volume, one goal of our work is to understand what types of support are most significant and in demand by young people themselves in their transition to adulthood and might carry a positive impact on the formation of the sense of agentic ability. Additionally, we try to think further about not planning as an actual type of planning strategy. As Peter Appleton (2019, p. 2) noted, young adults may or may not wish to plan in an explicit goal-oriented manner. First, emerging adulthood is regarded as an experimental period of life, characterized by exploration and instability (Arnett 2014). Second, for young people leaving care, multiple barriers may frustrate attempts to "get a life" (Pryce et al. 2017). Third, there is preliminary evidence that at least some young adults who are leaving care may be sceptical about future-oriented planning (Hung and Appleton 2016; Barratt et al. 2019). In comparison to other young adults—especially to those transitioning from ordinary family life—research has pointed out that the transition to adulthood of young adults transitioning from the care system is faster and more straightforward (Stein, 2006, p. 274). If more generally the literature on youth–adult transitions now speaks about yo-yo transitions, meaning that these transitions have become less linear, more complex and also reversible (Biggart and Walther 2006), these young adults usually need to transition to their independent life more rapidly and often with no place to return (Hiles et al. 2014, p. 1; also Lähde and Mölkänen 2021), i.e. with fewer chances to "make mistakes" in making decisions and choices concerning their later life (Kulmala et al. 2021a, p. 198).

Data and analysis

Our empirical data set consists of 43 biographical interviews with young care leavers, aged 17 to 31 (21.7 years on average) and either transitioning or having transitioned to independent living, in two regions of the Russian Federation of whom 26 are females and 17 males; respectively 26 live in our case study Region 1 and 17 in Region 2, as Table 9.1 illustrates.

Table 9.1 Studied care leavers, according to gender and region

	Region 1	Region 2	Total
Female	16	10	26
Male	10	7	17
Total	26	17	**43**

Both regions are located in the North-Western Federal District and they are of similar size in terms of population (500,000–1,000,000) and territory with around 80 per cent of urban population. In both regions around half of the people live in the capital. Young adults whom we interviewed in our study live in different places, including the capitals, small towns or villages. Both regions are industrial; at the same time, however, they are different in economic terms: Region 1 has significant natural resources and is significantly more developed than Region 2. According to the *Rosstat* data, Region 1 belongs to the first quarter of all Russian regions in terms of gross regional product per capita, while Region 2 is in the middle of the ranking. The average salary in Region 1 is 1.5 times higher than in Region 2.

All the interviewed young adults had lived in one or more forms of alternative care (children's home, foster family, children's village or a combination of these). Of our group, 15 grew up in an NGO-run children's village (*detskaya derevnya*), in which several foster families often live together with many children; 15 in residential institutions; and the rest had first been in residential institutions before being relocated to foster and guardianship families. Some of them had also been returned from foster family placements to residential institutions. For the sake of sensitivity, we do not name the studied Russian regions here, since in each of the studied regions, there is, for instance, only one children village per region. Moreover, all the people and organizations referred to and cited in this chapter have been anonymized.

We partly implemented our research through participatory research methods. Our academic research team interviewed 15 care leavers who were found through our collaboration with local child welfare NGOs.[3] Altogether six of the interviewed young care leavers, three in both case study regions, were recruited through consultation with the NGOs with whom we collaborated to peer-interview their fellows. These co-researchers conducted 28 interviews with their peers whom they contacted independently by themselves. These peer-interviewed young adults thus remained anonymous to us; as did their selection mechanisms. All interviews were conducted between December 2018 and October 2019 and they were recorded and transcribed. Additionally, after the peer-interviewing process, we arranged a study excursion to Finnish youth services for our co-researchers from Region 1, during which time we held a focus group discussion with them on their experiences in our project and what they learned most from the process and interviews they conducted. They were also welcomed to share their recommendations for the service system in question and decision-makers. We were in the middle of the arrangement of a similar excursion and focus group discussion to our co-researchers in Region 2, but unfortunately this was interrupted by the global COVID-19 crisis.

Since the topic is very sensitive and the interviewees and interviewers involved have most likely experienced severe hardships in their lives, we considered it of the utmost importance that the young people—both interviewers and interviewees—had a local focal point that they trusted and with which we have a confidential relationship. The local coordinator, employed by an NGO in both of the regions, assisted the interviewers. Her contact information was delivered to each of the interviewed persons, indicating that informants can turn to her with any issues or feelings related to the interviews. Any research with children or young adults involves ethical issues that need to be addressed, including concerns about possible exploitation, child protection, informed consent and gatekeeper issues (Törrönen et al. 2018). We have tried to be sensitive and reflective to any issues raised by the young adults involved in the process and spent time going through our research design and providing, alongside the needed research skills, training on numerous ethical issues, such as principles of confidentiality, anonymity and voluntary participation. These principles needed to be shared with everyone they interviewed and verbal consent was recorded at the beginning of the interview.

As our research focused on young people's agency, we found it impossible to carry out the research without the involvement of young people in the research process. As is usual with participatory research methods (e.g. Kilpatrick et al. 2007; Bradbury-Jones and Taylor 2015), we wanted to involve young people as active agents in our knowledge production and hopefully thus support their sense of agency. Our purpose has been to highlight young people as experts in their own life and the alternative care system in question through their personal experience, while providing some tools that can be useful in their work and study life: we, for instance, trained peer interviewers in qualitative interview and interaction skills at various points of the process[4] and gave certificates for their participation (Kulmala and Fomina 2019). We wanted to give young people a voice in understanding the forms of support that have been useful to them during and when leaving alternative care. Ultimately, we hope our research will bring improvements to these forms of support, which is why we emphasize the importance of collaboration with practitioners.

However, as self-reflected elsewhere (Kulmala and Fomina 2019), overall, our research design remained highly adult-led and could have obviously been more involving and participatory at the very first and final stages of the process. In other words, these young adults were not involved in designing our research questions, yet the used interview guide was elaborated with them. They have not been either involved in the empirical analysis, for instance, of this particular article. In the above-discussed focus group discussion, these young adults reported many kinds of benefits and learning processes they had gained during the process (Bradbury-Jones and Taylor 2015, pp. 163–165; Kilpatrick et al. 2007, pp. 367–368).

We have sought to overcome the asymmetric power relationships between the researchers and the researched (Kilpatrick et al. 2007; Bradbury-Jones and Taylor 2015). Anyone can recognize multilayered asymmetries in the situation where we, middle-class (and partly middle-aged) academically educated women, interview 20-year-old young people who have experienced situations leading to alternative care replacements. Through peer-to-peer talk, we have hoped to also open new perspectives on the studied issues. For example, similar experiences bring mutual understanding and language to interviews that perhaps allow for better communication and a more accurate reflection of young people's own thinking in our research materials (Törrönen et al. 2018; Kulmala and Fomina 2019).

Both groups of interviewers used the same interview guide. The interview questions followed a life-cyclic logic, including topics of birth family, placement in alternative, school and studies, work, housing, leisure time, close relationships, satisfaction with one's life and future plans. At all stages of life, we have tried to understand the involvement of the young person in the decision-making over their life and the kinds of support they have received around this (see Lähde and Mölkänen in this volume for a similar approach). Similarly, as Lähde and Mölkänen (this volume), we take a stance on what the authors call critical realism: we do understand that life is different as "lived" and "told" with which we acknowledge that "told" is a subjective interpretation by both, us and the interviewed young adults.

Analysis of the interviews was theory-led. Based on previous research on the planning strategies (long-term planning and not-to-plan) and modes of agency (life course and pragmatic agency) in the field of education (Kulmala et al. 2021a), we divided all the interviewed young adults into two (future) orientations (combined with information on gender, age, place of residence and experienced forms of alternative care) and then engaged in in-depth analysis of the factors interconnected with the orientations. Conceptually, this stage of the analysis was informed by the various theorizations on agency, while explanations concerning the observed orientations built on the analysis of the external contextual factors at the micro, meso and macro levels. The macro-level structural conditions include systems of social support, while the meso scale refers to local and regional infrastructures, including the availability of different forms of alternative care and support services. The micro level is more connected to individual life histories and relationships.

Planners, dreamers, copers, cynical "non-believers": future orientations among the studied young adults

We roughly divided all the interviewed young adults into two different dominant orientations to the future. In the first group, there are those oriented towards the future, those who planned and dreamed ahead, while the young adults in the second group did show no or little future orientation or even refused to plan.

Seventeen (five male, twelve female) of the 43 young adults belonged to the first group. Six grew up in a children's village, six in a residential children's home and five in a foster family. Nine lived in Region 1 and eight in Region 2. For the second group, we identified twenty (eleven male, nine female) young adults. Seven lived in a children's village, six grew up in a foster family and seven most of the time in a children's home. Fifteen of them lived in Region 1 and five in Region 2. With six young adults (five females, one male) it was difficult to name the dominant orientation but they somehow combined both two and are thus left out from the mentioned categories. Four of them grew up in foster families, the other two in children's homes; two lived in Region 1, while the rest four in Region 2. (See Tables 9.2 and 9.3.)

In both of the groups we identified four different—but obviously overlapping—subcategories: "long-term planning with strong self-efficacy", "dreaming-like planning", "unfeasible dreams", "planning with obstacles" in the first group. Under the second group there are: "no plans but current life satisfaction", "planning is not worth it", "no plans with survivalist self-reliance" and "no big plans but damn ordinary life".

Mostly planning took place around four main spheres, including education, work and family life and place of residence (cf. Massey et al. 2008). Almost all of the young people studied at the time of the interviews and accordingly made short-term plans for the graduation and postgraduation life, including taking advantage of the above-described right of the second free-of-charge educational programmes and unemployment benefit guaranteed for care leavers by the law. As the system heavily directs care leavers to study (also Kulmala et al. 2021a), it is no surprise that only two young men had decided not to continue in education: one of them had never liked school, while the other one had a life situation that required him to find a job instead of a school. Almost all of them thought that studying would help them find a job, but six interviewed young adults did not quite know what job that

Table 9.2 Studied care leavers, according to orientation group (OG) and region

	OG 1	OG 2	N/A
Region 1	9	15	2
Region 2	8	5	4
Total	**17**	**20**	**6**

Table 9.3 Studied care leavers, according to orientation group (OG) and gender

	OG 1	OG 2	N/A
Male	5	11	1
Female	12	9	5
Total	**17**	**20**	**6**

would be and how to find it. Three young adults saw it as important to finish their studies in order to have education, but at the same time they did not necessarily believe that it would help them in finding a good job; two thought that the skills they are learning will not be useful in the future. At the same time, 13 could name the sphere where they would like to work. The rest were not only able to envisage their future job but had also concrete and detailed plans on how to get there. In regards to longer-term plans, five young adults wanted to have a well-paid job in the future, while others connected wellbeing with "finding a good place" more generally in their lives.

   One of the key topics was family making. Two of the young women, who had a relatively long and stable partnership, were planning to get married in the nearest future. 24 planned children (12 female, 12 male), 14 (5 female, 9 male) of them in a more distant future, when they have become independent and had stable life (in terms of job and income). Most of the young women clearly postponed the decision about having a child by explaining the decision—not only as a rational planning related to the stability and career—but also as an aspiration to focus on oneself to find and better understand their partners. In Region 1, three women had a child at the moment of interview and one young man was expecting one with his partner. In Region 2, four women had children and none of the male participants had children.

   Our interview guide specifically included a question about how the interviewees see their life within the next five to ten years. Most imagined their life positively: having a well-paid job with family and friends. Besides, they talked about some individual plans, such as starting to do sports, and learning languages. Several wanted to move to another city, usually St Petersburg or Moscow, or even abroad, mainly to Scandinavia or the USA. Moving to other parts of Russia was connected to better possibilities in the labour market, while emigrating to Finland and Norway was associated with societal security. Next, we turn to a more detailed analysis of these two categories to understand what kinds of other elements they interconnect with.

Young adults with future orientation

Long-term planning with strong self-efficacy

We identified eight young adults (seven female, one male) who engaged in strong future orientation, including long-term, life-course-type planning with strong beliefs on one's own ability to influence one's life course (self-efficacy). This mode was typically related to the fact that these young adults had someone who believed in them, with whom they could talk about their plans and who supported their choices. In other words, they had a trusted adult with whom they could discuss choices, as one young woman described the process with her foster mom concerning her education:

> Let's say I wanted to become a designer. Mom would tell me: "In XXX [the name of the town], you'd better become a cook". I would say to her:

"Mom, here [the town] every third person is a cook, there is no develop-ment in that field". She says: "Well, right, yes, good. Go to XXX [the capital city of the region], only you need to be careful" [...] Let's say I want to get higher education. I say: "Mom, I want to be a social worker to work with youngsters ... She would say: "Well, yes, social work with the youth is your [field], it definitely suits you." She offers some ideas of hers, but she is receptive to mine as well.

(F, 21, #18, R1)

Or, as a young man explained his high-level trust to close people if things would not go as planned: "If I really did bad, I could turn to my friends, I could turn to my [foster] parents, I could turn to my godmother or someone else. I always have had people I could turn to ..." (M, 21, #4, R1). In terms of agency, such future orientation combined with the ability to plan ahead can obviously be connected to life course agency, but it can be also conceptual-ized as "shared agency" with acting and making decisions jointly together and getting support for one's choices.

Often such life course agency is connected with strong identity construc-tion in terms of future profession. These young adults often knew well what they want to be and made logical educational and related decisions, as in Kulmala et al. (2021a). One young woman (F, 22, #39, R1) had considered all the options in a branching plan "development of my life" that she had drawn; another (F, 21, #23, R1) had a foster mother who had supported her educa-tion since her early childhood and, as a result, ended up making several stra-tegic choices in order to become a social worker.

As above, identity agency—a more or less conscious act to make use of one's identity in compliance to social norms typically related to this iden-tity—can be positive in that one has a clear image of what (s)he wants to be as a 'grown up'. Yet, this mode more than once was connected with what we label as 'negative identity agency': that which a young adult does not want to become, as described by a young woman concerning upbringing children:

I don't say I don't want children, I just want them not to have the kind of childhood I had ... if my child asks me: "Mom, I want an iPhone 6, for 25,000 thousand". I don't want to answer: "Sorry, we don't have money, we can only eat buckwheat". Instead [I want that] I'll go and buy that phone. Well, I want to stand on my own feet, I want to have of stability in my job, not just any job, but a prestigious job, or I want to start my own business...

(F, 21, #24, R1)

Negative identity agency is usually connected with dysfunctional behaviour on the part of birth parents (e.g. alcohol/drug abuse) and/or a constant lack of material things (see also Shpakovskaya and Chernova, 2021). In both ways, identity agency can serve for a certain degree of strategic resistance (cf. Lister 2004, pp. 140–141) and thus as a resource to build plans.

Dreaming-like planning

Other respondents were future-oriented but without identifying concrete plans, let alone efforts to make them happen. These young adults engaged in planning in a more dream-like manner. They had in common a positive tone when talking about the present and future and showed more or less self-confidence in terms of having an impact over their own life, but it remained unclear for us what has been or is to be done to make those dreams come through. One young man who had succeeded in many of his aspirations thought his plans to move might become real:

> Listen, we want to get the hell out of here. Shit, to somewhere. Well Anya [the name of his wife changed] wants to go to the USA. I don't know how to get there, of course, but we simply have a goal to move away from here. Simply to get away. In any case, we just need to move somewhere from here ... Hell. I think that a chance will open quite soon. I mean some kind of a purpose [of life]. Because, with shit like this, one can't continue ... to do things that don't please you. I want [to live] maximally. To do what one wants to do ... I don't know. I don't have any strateg[y] ... I am not any [strategic] planner. It is Anya who plans.
>
> (M, 29, #43, R1)

Or as expressed in the words of one young woman, who showed neither strategic planning nor a strong sense of agentic ability but who, alongside her expectation for many positive things "to happen", seemed to have much trust in the future when she dreamed about her future family:

> [I will live] with a baby doll, a little baby doll who has a dark blue baby carriage made from organic leather with a price of 30,000 [Roubles; equivalent to 385 euro] which I really want! The baby's dad works will have a permanent job." (...) I will work with my daughter! My workplace is maternity leave. I will give birth to another baby girl. (...) We will live, I hope, in our own apartment, well, in a new one.
>
> (F, 25, #5, R2)

Similarly, as future orientation and planning contribute to resilience (Hitlin and Kirkpatrick Johnson 2015; Appleton 2019), we assume that a positive life attitude combined with daydreaming is a resource to a certain extent, especially when compared to the young adults discussed below, who regarded their future with cynicism. .

Unfeasible dreams

We also witnessed dreams that, in the context of the rest of the talk, had no realistic prospect of coming true. For example, in one interview a young mother spoke of her plans to move with her new boyfriend to St Petersburg, or even the USA:

I met my boyfriend quite recently … He will come to take up me and the baby to St Petersburg. (…) I will move to live in St Petersburg. I will most likely study there. I haven't chosen my profession yet but I would really like to [study]. I don't know [why St Petersburg], I like the city. Or another big city … Well, I have a dream, simply a dream to study in the US. (…) In five to ten years, we will be on vacation in the US. My child is already grown-up. Well, we will live in St Petersburg, everything is fine with us.

(F, 19, #22, R1)

There were many inconsistencies in her talk about the present and the future. As in her case, this type of dreaming seemed typically to be connected to a rather low level of overall life management and self-regulation. As in her case, she had hardly finalized anything she had started. "Unfeasible plans" is obviously our interpretation; the young adults did not use such terms. Our interpretation of unfeasibility is based on the overall context of their narratives and intonations and how they spoke about their future and plans (sometimes even with irony and cynicism, as below). In this orientation subcategory, their narrations were typically characterized with inconsistencies (e.g. a new-born baby soon grows up, as above) and obviously unrelated issues turned out to be related in these narratives. Moreover, these young adults narrated little ability to consciously control the circumstances on which their future depends. As we discuss below, such orientation is not considered as any individual failure, but it logically interconnects with many past experiences and structural constraints.

Planning with constant obstacles

We identified several young adults who made plans and engaged with efforts to make those plans happen, yet the external circumstances repeatedly thwarted those plans. One young man (M, 21, #17, R2) wanted to enrol in higher education as a computer programmer, but *"likely, nothing will work out because of the [required] high scores and paid education"*, which is why he decided to go to a college in a small town instead of the capital of the region where he really wanted to go: *"I didn't have a good enough diploma so I had to go to X [name of the small town]."* Yet, again, he wanted to have an apartment in the capital; but due to state policy, however, he will receive one in another small town. Yet, he kept planning. Another example was that of a young woman (F, 21, #21, R1) who wanted to become an animal attendant, but could not do so as there were no such schools in her region. She changed her plan, now aiming to become a car mechanic, but could not do that either because girls were not accepted onto that programme.

Such situations are ideal-typical cases of restricted, or constrained agency. These young adults are by no means passive objects without agency, but they cannot act freely either (cf. Viuhko 2020, 45): they are constrained by the external circumstances within which they act. Some people are more resilient and stay optimistic enough to make another plan, while others quite logically lose their faith in any planning.

Young care leavers with no or little future orientation

As in other studies (Hung and Appleton 2016; Appleton 2019; Barratt et al. 2019), a number of the care leavers in our study, twenty out of 43 (nine female, eleven male), expressed strong intentions not to plan for the future.

No plans but current life satisfaction

Unwillingness to plan was sometimes connected to more or less satisfied and optimistic attitude over one's life: "*Everything will be fine*" (M, 21, #8, R1), as one young man saw his future life. What these young adults shared in common was low future orientation; at the same time, however, they engaged with a somewhat optimistic mode of life. They often saw that one cannot plan too much for the future, but that anyhow things tend to go well in the end. They did not express strong self-efficacy in terms of being able to have an impact on their life course but since they were pretty satisfied with their current life, they had no reason to think that something will go wrong. Similarly, as above, such an optimistic attitude can be a sign of a certain level of resourcefulness and resilience.

Planning is not worth it

However, more typically unwillingness to plan was expressed through a rejection of future planning due to a disbelief in having control over one's life. There is no sense in planning because everything can just change, as several young adults stated: "Who knows how life will turn out?" (F, 22 #20, R1); "I don't know about my plans so I don't start guessing for the future" (F, 25, #6, R2); "I don't plan that much ahead, I can't respond to this question [concerning the future plans]" (F, 21, #33, R1).

These people share in common is a weak sense of mastery over things in their lives. As the above-discussed "unfeasible plans", this orientation was often connected with little life management and self-regulation more generally, as the quotation from one young woman illustrates well:

> I went to a grocery store to buy toilet paper. I came out with a full shopping bag with (a price of) 1,500 [Roubles, equivalent to 20€] ... I came home and unpacked the bag. No food. I was like an autopilot. And I thought, this is exactly what life will [continue to] be.
>
> (F, 25, #12, R2)

The rejection of planning was sometimes tightly connected with irony or cynicism to talk about the future, as one male care leaver said:

> [I would like to be able to do] programming, speak ten languages, uh, to be able to fly, construct computer networks ... Within five years, excellent. This is the plan ... Let's say in, uh, I will live somewhere ... close to

the equator ... Somewhere in Madagascar. There. Friends, family, there. A child ... [laughs]

(M, 22, #9, R2)

Cynicism can be interpreted though as an expression of everyday resistance (Lister 2004, 144): it again shows agency; instead of passivity, it is an active act of the rejection of planning, which is perhaps a more logical act if one has not experienced any positive worth of planning. It is thus more about one doing what one has most reason to do (cf. Appleton 2019, pp. 7–8).

No plans with survivalist self-reliance

In line with the findings of Appleton (2019) and Pryce et al. (2017), we witnessed "survivalist self-reliance" among those who refused to plan. These people felt responsible for their own development and safety and expressed mistrust towards others as potential sources of support (Pryce et al. 2017). Such a lack of trust in anyone's help and one being on his own is well illustrated by a young man describing his experience of being in a children's home: "I understood it all at once ... if you don't do anything by yourself, nobody will give you anything" (M, 31, #2, R2). As Hung and Appleton found (2016, p. 43), these young adults see life being a day-to-day survival in which self-reliance was essential. As in Pryce et al. (2017, p. 318), this pairs with negative associations of the help from the "system" and obviously connects with earlier experiences of untrustworthy parents. With early unmet needs for help, ambivalence about asking for help is a logical choice.

No big plans but "damn, ordinary life"

Several young adults refused to plan but 'just' wanted some stability. We could sense fatigue in the face of many kinds of challenges that they had encountered. They did not ask much, they did not plan big; they just wanted a quiet, ordinary life. As one male put it: "Well, probably regular work where I will go every day ... My own apartment. Well, damn it, ordinary life" (M, 21, #1, R2). In contrast to those with optimism that life will go well whatever happens, what is common to this subgroup is a somewhat pessimistic attitude: they wanted "just a normal life" without faith that it will come. These people, are not completely lacking agency. This kind of coping, at a very minimum, is "an active process of juggling, piecing together and going without" (Lister 2004, p. 133).

The future orientation of the young adults in our research varied from the complete refusal to plan to systematic planning and judgement of the different options. The observed stances varied from no trust in one's own agentic ability to a strong sense of mastery. As David McCrone (1994, pp. 70–80) emphasized, the distinction between "non-planners" who "get by" on a day-to-day basis and "planners" who "make out" through the deployment of longer-term strategies is a very thin line. Also getting by might require lots of competence to run through daily routines, which might be so burdening that it makes it difficult to think or act strategically (cf. Chamberlayne and Rustin 1999). As concluded by Kulmala et al. (2021a), the (in)ability to make long-term plans is not a "success" or "failure" of the young adult themselves, as there are external factors that affect the modes and orientation.

Facilitations and constraints of future-oriented agency at the macro, meso and micro levels

There are various levels of constraints that challenge the transition to adulthood for this particular group of young people (also Pryce et al. 2017). Next, we aim to understand the factors at the macro-, meso- and micro-levels that either facilitate or constrain the future-orientated agency.

The macro level

Russian state policy supports this specific group of young adults with a wide spectrum of benefits, which shape their plans. Of these measures, the right of secondary vocational and university degrees up to the age of 23 is the most significant. As a result of this policy, most care leavers choose to study (Kulmala et al. 2021a). Our analysis here also shows that this right determines the direction of the short-term plans of the care leavers in our research, who apparently and understandably want to take the advantage of the benefits provided by the state:

> I want to study because up to 23 years old I can enter [school] free-of-charge. I have such a chance. I don't want to waste it. Why would one go studying later and have to pay? If there is such opportunity, I don't understand why would someone who is 21 years old want to sit in the office somewhere, I don't know, to work (...)
>
> (F, 21, #18, R1)

Generally, the opportunity to study and the fact that most care leavers study, which is not anything obvious in the international comparison, can be considered as an enabling macro-level factor. As shown by Kulmala et al. (2021a), however, there are also many constraints. One cannot speak about free choice, but societal norms (e.g. stigma), regional-local infrastructure (labour market, availability of educational institutions) and the lack of information and individual counselling and support might heavily limit this choice (also Lähde and Mölkänen, this volume). Young adults also tend to take the advantage of the earlier-discussed unemployment benefit. It seems that there is much less support and guidance to navigate the job market to find a place to study (also Kulmala et al. 2021a).

State housing policy offering a free apartment of one's own is generous and quite unusual in international comparison. Yet there are many shortages and regional differences in implementing this right. It can be viewed as significant material capital, but also as shaping long-term plans. For example, having an apartment in a certain town and region, it is more difficult (to plan) to move somewhere else to study or work even if one wished so. The huge differences in income and prices across Russia do not always allow young people to sell apartments profitably in their home region in order to move to another one: for instance, given the price of an apartment in a Karelian municipality one can hardly buy any place to live in Saint Petersburg. Additionally, according to the law, care leavers can sell the apartment only after five years of its receipt. Anyhow, many plan or dream about moving to bigger cities or even abroad. Due to the location of North-West Russia, young adults were familiar with the Scandinavian countries and some showed willingness to move there: "Someday perhaps I will move somewhere ... Well, to Norway, Finland ... Scandinavia ... According to the statistics, the happiest people live there ..." (F, 21, #18, R1). These plans were, however, as discussed above, often somewhat presented as unfeasible or accompanied by cynicism.

As Ruth Lister (2004, p. 10) aptly argued, cultural meanings and societal norms create the context within which people exercise their agency. This becomes especially significant when we speak about people in vulnerable situations (cf. Hitlin and Elder Jr 2007, p. 177). As discussed earlier, children deprived of parental care bear strong stigma in Russian society (Khlinovskaya-Rockhill 2010; Iarskaia-Smirnova et al. 2021) which can be considered as a macro-level constraint: for instance, the stereotypical image of these children as bad students obviously affects their educational choice - and advice (cf. Kulmala et al. 2021a). Yet, as we showed in this chapter, sometimes negative identity agency is exercised to resist the stereotypes and expectations of bad outcomes. In a similar way, also stereotypical understanding of gender differences can be viewed as a structural constraint (Kulmala et al. 2021a).

It is important to note that all the benefits are dependent on the status of being "deprived of parental care". Even if much-needed, the benefits—and especially their implementation—might be based on certain stereotypes and thereby reinforce them (Lister 2004, pp. 101–102). Many of the young adults indeed appreciated the benefits but found them stigmatizing at the same time, as well illustrated by a quotation from the focus group discussion conducted by us:

ASYA: I never took any of the social benefits that are offered. I'm so embarrassed to take them. For instance, in college they gave bed sheets ... Yes, Olya, what did they give? Olya brought them to me.

OLYA: Yes.

ASYA: Basically, we were not friends then but I said to her: "Jesus, Olya, I won't go there, I won't carry that sack throughout college." And I didn't pick them up for myself, of all those huge bags I only took one blanket. (...)

ANYA (INTERVIEWER): Why?

ASYA: I don't know, it sucks to take help. I never ate in the college cafeteria for free, I always bought food with my own money.

The public image of being a marginalized and stigmatized group limits opportunities generally while also affecting the subjective perception of one's own available opportunities and more widely of mastery over one's own life (cf. Hitlin and Kirkpatrick Johnson 2015). As Lister (2004) argued, some identities are more collectively mobilizable than others: for instance, other people might easily find a common case, while poor people, stigmatized in many contexts, do not. Yet, as our study shows, active resistance to stigma can serve as a resource.

The meso level

Even if the massive child welfare reform uniformed in principle the public alternative care system throughout the country, in practice Russian regions, which are in charge of implementing the child welfare policy (Kulmala et al., 2021c), are obviously very different when it comes to resources (Kainu et al. 2017). The regions vary, for instance, when it comes to the labour and housing market or educational sphere. The available options obviously impact the future plans of young people. We would argue the most significant meso-level factor is the availability and use of different forms of alternative care also outside the public sector in the region in question.

Both of the studied regions have a long history of transnational collaboration with NGOs, including in the sphere of child welfare. Yet Region 1 is clearly more resourced and developed in terms of family forms of alternative care. There is one NGO, in particular, established after transnational collaboration in the early 1990s, which provides family-based alternative care in a children's village and has exceptionally developed aftercare services for the care leavers. There is also another NGO with a well-resourced programme of individual mentors to care leavers. Many of the care leavers in our research had benefited from these. Through cross-sectoral collaboration, such developed expertise on family-based care arrangements and importance of aftercare support also transfers into now developing public services. Also, in Region 2 there are developed child welfare NGOs, but their work has been more focused on material assistance of children in residential care instead of the wider reconstruction of the care system itself. The care leavers in our research from this region reported occasional connections with the NGOs. Especially in this region, it seems that both public and third sector support for care leavers remains unsystematic and even random depending on "lucky circumstances" and "good people". Yet, as the services in Region 1 are mostly provided on a project and grant basis by NGOs, their coverage is limited and their continuity uncertain.

In a study on the educational choice by Kulmala et al. (2021a), it became clear that that young adults who live in residential institutions are less informed about their options, while NGO-run children's villages, in particular, did better through their emphasis on individual support, counselling and encouragement—which brings us to the importance of the micro-level environment.

The micro level

The micro level, the level of everyday interactions, proved to the most significant factor for the development of the future-orientated self-efficacy among the young adults in our research. We found that from those eight who were included in the category of "long term planning with strong self-efficacy", four grew up in children's villages, two in foster families, and only two in children's homes but, essentially, with the support of significant adults, either someone from a children home and/or a birth parent. Seven out of eight care leavers talked about the important influence of their foster parents, or some other significant adults, in the significant choices they made in their life. Thus, they have had someone who has listened to their needs and wishes and in the end believed in them. In other words, they have had a trusted one to negotiate and practice shared agency. This person does not have to be a family member; the important point is that a young person trusts someone to discuss their future plans (Pinkerton and Rooney 2014). This significant finding is very much the same as Lähde and Mölkänen reported in this volume based on their study concerning young adults who have sought support from certain NGO-based services in the Finnish context. This strongly emphasizes the importance to guarantee that also the young people deprived of parental care would have long-standing, robust relationships with adults.

On the other side of the coin, those young adults who refuse to plan come with various backgrounds in alternative care. Many have lived in residential institutions, but some of them also have grown up in foster families or villages. Each of them, obviously, has faced severe hardships with their birth families and even later with foster families. Their social relationships have been disrupted. As Appleton (2019) pointed out, young people with a history of maltreatment and alternative care usually have experienced flouting and violation of rational and planning norms by significant others—a birth parent, sometimes a foster parent, and sometimes a public service provider (also Pryce et al. 2017). Life has most likely brought up endless occasions that one had not planned or even wanted to happen. Perhaps inconsistency is the only stable element. Based on this, rejecting planning and self-reliance can be considered as a consistent, logical continuum striving from the past life history (see also Lähde and Mölkänen, this volume). One has perhaps ended up "doing what (s)he has most reason to do" (Appleton 2019, p. 7).

Based on our analysis (also Lähde and Mölkänen, this volume), it is evident that future-oriented agency combined with present understanding of one's own agentic abilities builds on past experiences. Having a trusted someone listening and believing in you most likely positively contributes to self-esteem and positive identity construction and further to self-efficacy. As illustrated, also identity agency matters: the more clearly one knows who (s) he wants—or does not want—to be or become, the better abilities one has in life course planning and enjoying an overall sense of control over one's life (cf. Côté 2016, p. 31).

Conclusion

Our investigation shows that agency, self-efficacy and projectivity are obviously interconnected. Given the fact that these all contribute to resilience to cope with challenges, it is indeed important to reveal the conditions that promote the development of these abilities among this particular group of young adults who have faced severe hardships in their life. We conclude that the agency of young adults in our study is restricted in many ways by external factors. On the other hand, our study shows that future-oriented agency could be supported by individual support by a significant adult (whomever this is).

In Russia, the macro-level social policy substantially supports the children deprived of parental care. At the same time, these policy measures might stigmatize their targets and limit their future choices. It is important indeed that state policy directs these young adults to study, but, for instance, by supporting certain career paths, for instance vocational education for certain professions, which narrows the choice and agency of these young people (Chernova and Shpakovskaia 2020; Kulmala et al. 2021a). Similarly, the (generous) housing policy leads to certain residential pathways. Even if the Russian system can be considered good and generous in terms of material support, it simultaneously fails to provide much-needed individual (emotional) support to young adults, as our study clearly points out (also Kulmala et al. 2021a). As the study by Lähde and Mölkänen (this volume) shows, this can also be the case in other contexts.

Not a surprising but highly important finding of our study (as with Lähde and Mölkänen, this volume) is that micro-level supportive and trustworthy relationships essentially matter to the development of the sense of control and ability to influence one's life course and future orientation. As shown, the young adults in our study were almost equally divided into two orientations of planning (n = 17) and not-planning (n = 20) at various extents, but those who engaged in life course planning with strong self-efficacy reported longer-lasting relationships with a significant, trustworthy adult, be it a foster or birth parent, individual mentor or pedagogue. Moreover, most of them grew up in a foster family in a children's village with many kinds of support programmes and services. This allows us to conclude that a family-like environment with systematic approach to aftercare support is something to be developed further. Ongoing reforms in Russia have taken many steps in the right direction, but with many pitfalls that hamper the formation of individual support. As has been argued elsewhere (Jappinen and Kulmala 2021), the good intentions of these reforms result in many unintended consequences. In the current political environment, quantitative measurements to show good results in numbers become more important than changes in the quality of care, which, in turn, leads to situations when some other rights—for instance those of foster parents—override the rights of children.

Our study suggests that without strong individual support at the micro level, the majority of the care leavers in our study fail to plan and build their desired future. One can, of course, argue that the "desired future" is heavily directed by expectations of a normative path of becoming a 'proper' citizen, who is integrated to study and work life (Furlong 2012). On the other hand, research has shown that setting and pursuing goals is particularly pertinent during adolescence when establishing identity is of fundamental importance. Planning and goal pursuit serves not only as a self-directing and self-defining process, but is also affiliated—when successful—with positive affect and well-being (Massey et al. 2008, p. 422). This is also the case with the sense of mastery connected to resilience (Hitlin and Kirkpatrick Johnson 2015). This is why we view the ability to plan and the sense of mastery as important resources for care leavers, who often transition to their independent living in a more vulnerable situation than other young people. In our study, we found that the ability to plan goes hand in hand with identity agency and suggest, similarly as Pryce et al. (2017, p. 318), that future research on care leavers should focus on identity and self-concept as central issues to understand resilience in adulthood. Future research on care leavers should focus on identity and self-concept as central issues to understand resilience in adulthood. Moreover, in our study the vast majority of young adults with strong future orientation and self-efficacy were women. Obviously, in the future more focus is needed on this gender difference.

Even if our study concerned the Russian Arctic, its most important argument concerning the importance of individual support by a significant adult is obviously global in scope. As Lähde and Mölkänen showed in their chapter in this volume, this is the case also in Finland and we can assume that it is the case everywhere, regardless of geographical location. We also expect that many other findings are similar in other parts of Russia. Young care leavers in all the Russian regions are supported by similar state social policies, but still may face shortages in individual and emotional support. Nonetheless, some regions are more equipped, for instance, with NGOs, which might have a significant role for the development of more individually oriented alternative and aftercare services for young care leavers.

Notes

1 By alternative care we refer to all forms of out-of-home care for children deprived of parental care such as residential children's homes and different types of foster families (see e.g. Kulmala et al., 2021c)

2 These interviews are a part of the larger data set collected in two separate but interrelated research projects. One was led by Meri Kulmala on 'A Child's Right to a Family: Deinstitutionalization of Child Welfare in Putin's Russia' (2016–2020), funded by the Academy of Finland (No. 295554), University of Helsinki (ref. 412/51/2015) and Kone Foundation (cd276a and df3277). The other focused on youth wellbeing in the Arctic: 'Live, Work or Leave? Youth—wellbeing and the viability of (post) extractive Arctic industrial cities in Finland and Russia (2018–2020)', funded by the Academy of Finland and Russian Academy of Science (AKA No. 314471, RFBR No. 1859–11001).

3 We conducted 43 interviews with representatives of Russian child welfare NGOs in the project led by Dr. Kulmala (see endnote 1). Additionally, we participated in and arranged five research-practice seminars with mainly Russian child welfare street-level practitioners, including NGOs, during which we engaged in close dialogue with these practitioners. (See more e.g. Kulmala et al. 2021d). First interviews in Region 1 were conducted by Zhanna Chernova, Meri Kulmala and Anna Tarasenko, while Anna Fomina took all the non-peer-interviews in Region 2. As the interviews in Region 1 were biased with a particular form of alternative care (Kulmala et al. 2021a), Fomina returned there to conduct a few more interviews with care leavers having background in residential care.

4 We trained them face-to-face before and online after the pilot peer-interviews as well as in the middle of the interviewing process. These methodological issues were also raised in the focus group discussions at the end of the data collection process.

References

Abramov, R.N., Antonova, K.A., Il'in, A.V., Grach, E.A., Lyubarskiy, G.Y. and Chernova, Z.V. (2016) *Trayektorii sotsial'noy i professional'noy adaptatsii vypusknikov detskikh domov v Rossii (obzor issledovatel'skogo otcheta)* [Trajectories of social adaptation of care leavers of children's homes in Russia (review of the research report)]. Moskva: SB Grupp.

Appleton, P. (2019) 'Anchors for deliberation and shared deliberation: Understanding planning in young adults transitioning from out-of-home care', *Qualitative Social Work*, 19(5–6), pp. 1–17. doi: https://doi.org/10.1177/1473325019869810.

Arnett, J. (2014) *Emerging adulthood: The winding road from the late teens through the twenties*. 2nd edn. Oxford: Oxford University Press.

Barratt, C., Appleton, P. and Pearson M. (2019) 'Exploring internal conversations to understand the experience of young adults transitioning out of care'. *Journal of Youth Studies*, 23(7), pp. 869–885. doi: https://doi.org/10.1080/13676261.2019.1645310.

Biggart, A. and Walther, A. (2006) 'Coping with yo-yo-transitions: Young adults struggle for support, between family and state in comparative perspective', in Ruspini, E. and Leccardi, C. (eds.), *A new youth? Young people, generations and family life*. London and New York: Routledge, pp. 41–62.

Biryukova, S. and Makarentseva, A. (2021) 'Statistics on the deinstitutionalization of child welfare in Russia', in Kulmala et al. (eds.), *Reforming child welfare in the post-Soviet space: Institutional change in Russia*. London and New York: Routledge, pp. 23–46.

Bradbury-Jones, C. and Taylor, J. (2015) 'Engaging with children as co-researchers: Challenges, counter challenges and solutions', *International Journal of Social Research Methodology, 18*(3), pp. 161–173. doi: https://doi.org/10.1080/13645579.2013.864589.

Chamberlayne, P. and Rustin, M. (1999) *From biography to social policy*. London: Centre for Biography in Social Policy.Brat.

Chernova, Z. and Shpakovskaia, L. (2020) 'Vverkh na lesnitse, vedushchei vniz: Ideologiia vozmozhnostei i ogranicheniia obrazovatelnogo vybora vypusknikov detskih domov' [Up to the staircase that take down: ideology of opportunities and constraints of educational choice of care leavers of children's homes in Russia].

Zhurnal sotsiologii i sotsial'noi antropologii, XXIII/2, pp. 104–129. doi: https://doi. org/10.31119/jssa.2020.23.2.5.

Côté, J. (2016) *The identity capital model: A handbook of theory, methods, and findings.* Ontario, Canada: The University of Western Ontario.

Decree #481 Postanovlenie Pravitel'stva Rossiiskoi Federatsii ot 24 maia 2014 g. N 481 g. (2014) *Moskva O deiatel'nosti organizatsii dlia detei-sirot i detei, ostavshihsia bez popecheniia roditelei, i ob ustroistve v nikh detei, ostavshihsia bez popecheniia roditelei'* [Decree of the Government of the Russian Federation of May 24, 2014 No. 481 'On the performance of organizations for orphaned children and children without parental care, and on placement of children in these organizations]. Available at: http://static.government.ru/media/files/41d4e0dc986dd6284920.pdf (Accessed: March 30 2020).

Emirbayer, M. and Mische, A. (1998) 'What is agency?', *The American Journal of Sociology*, *103*(4), pp. 962–1023. doi: https://doi.org/10.1086/231294.

Furlong, A. (2012) *Youth studies: An introduction.* London and New York: Routledge.

Hiles, D., Moss, D., Thorne, L., Wright, J. and Dallos, R. (2014) 'So what am I?— Multiple perspectives on young people's experience of leaving care', *Children and Youth Services Review*, *41*, pp. 1–15. doi: https://doi.org/10.1016/j.childyouth.2014. 03.007.

Hitlin, S., and Elder Jr, G.H. (2007) 'Time, self, and the curiously abstract concept of agency', *Sociological Theory*, *25*(2), pp. 170–191. doi: https://doi.org/10.1111/ j.1467-9558.2007.00303.x.

Hitlin, S. and Kirkpatrick Johnson, M. (2015) 'Reconceptualizing agency within the life course: The power of looking ahead', *American Journal of Sociology*, *120*(5), pp. 1429–1472. doi: https://doi.org/10.1086/681216.

Hung, I. and Appleton. P. (2016) 'To plan or not to plan: The internal conversations of young people leaving care', *Qualitative Social Work*, *15*(1), pp. 35–54. doi: https://doi.org/10.1177%2F1473325015577408.

Iarskaia-Smirnova, E. Kosova, O. and Kononenko, R. (2021) 'The 'last-minute children': Where did they come from, where will they go? Media portrayals of children deprived of parental care', in Kulmala et al. (eds.), *Reforming child welfare in the post-Soviet space: Institutional change in Russia*. London and New York: Routledge, pp. 47–67.

Jappinen, M. and Kulmala, M. (2021) 'One has to stop chasing the numbers': the unintended consequences of Russian child welfare reforms', in Kulmala et al. (eds.) *Reforming child welfare in the post-Soviet space: Institutional change in Russia.* London and New York: Routledge, pp. 118–140.

Kainu, M., Kulmala, M., Nikula, J. and Kivinen, M. (2017) 'The Russian welfare state system: With special reference to regional inequality', in Aspalter, C. (ed.) *The Routledge International Handbook to Welfare State Systems*. London and New York: Routledge, pp. 291–316.

Khlinovskaya-Rockhill, E. (2010) *Lost to the State. Family discontinuity, social orphanhood and residential care in the Russian Far East.* New York: Berghahn Books.

Kilpatrick, R., McCartan, C., McAlister S. and McKeown, P. (2007) 'If I am brutally honest research has never appealed to me ...: The problems and successes of a peer research project', *Educational Action Research*, *15*(3), pp. 351–369. doi: https://doi. org/10.1080/09650790701514291.

Kulmala, M., Chernova, Z. and Fomina, A. (2021a) 'Young adults leaving care in North-West Russia: Agency and educational choice', in Kulmala et al. (eds).

Reforming child welfare in the post-Soviet space: Institutional change in Russia. London and New York: Routledge, pp. 198–220.

Kulmala, M. and Fomina, A. (2019) 'Luoteisvenäläiset nuoret kanssa- ja vertaistutki-joina [North-Western Russian young adults as co- and peer-researchers]', *Idäntutkimus [The Finnish Review of East European Studies]*, *26*(4), pp. 96–102. Available at: https://journal.fi/idantutkimus/article/view/88851 (Accessed: May 27 2020).

Kulmala, M., Jäppinen, M., Tarasenko, A. and Pivovarova, A. (2021b) *Reforming child welfare in the post-Soviet space: Institutional change in Russia.* London and New York: Routledge.

Kulmala, M., Jäppinen, M., Tarasenko, A. and Pivovarova, A. (2021c) 'Introduction: Russian child welfare reform and institutional change', in Kulmala et al. (eds), *Reforming child welfare in the post-Soviet space: Institutional change in Russia.* London and New York: Routledge, pp. 3–19.

Kulmala, M., Shpakovskaya, L., and Chernova, Z. (2021d) 'Ideal organisation of Russia's out-of-home care: Child welfare reform as a battle over resources and rec-ognition', in Kulmala et al. (eds), *Reforming child welfare in the post-Soviet space: Institutional change in Russia.* London and New York: Routledge, pp. 71–94.

Lähde, M. and Mölkänen, J. (this volume) 'The quest for independent living in Finland: Youth shelter as a critical moment in young adults' life courses', in Stammler, F. and Toivanen, R. (eds) *Young people, wellbeing and placemaking in the Arctic.* London: Routledge, pp. 170–195.

Lister, R. (2004) *Poverty.* Cambridge: Polity Press.

Massey, E. K., Gebhardt, W. A., & Garnefski, N. (2008) 'Adolescent goal content and pursuit: A review of the literature from the past 16 years'. *Developmental Review,* *28*(4), pp. 421–460. doi: https://doi.org/10.1016/j.dr.2008.03.002.

McCrone, D. (1994) 'Getting by and making out in Kirkcaldy', in Anderson, M., Bechhofer, F. and Gershuny, J. (eds.), *The Social and Political Economy of the Household.* Oxford: Oxford University Press.

Nurmi, J.-E. (1991) 'How do adolescents see their future? A review of the develop-ment of future orientation and planning', *Developmental Review, 11*(1), pp. 1–59. https://doi.org/10.1016/0273-2297(91)90002-6.

Pinkerton, J., Rooney, C. (2014) 'Care leavers' experiences of transition and turning points: Findings from a biographical narrative study', *Social Work & Society, 12*(1), pp. 1–12. Available at: https://www.socwork.net/sws/article/view/389/730 (Accessed May 29 2020).

Pryce, J., Napolitano, L. and Samuels, G.M. (2017) 'Transition to adulthood of for-mer foster youth: Multilevel challenges to the help-seeking process', *Emerging Adulthood, 5*(5), pp. 311–321. doi: https://doi.org/10.1177%2F2167696816685231.

Shpakovskaya L. & Chernova Z. (2021) 'The successful transition to foster care: The child's perspective', in Kulmala, M., Jäppinen, M., Tarasenko, A. & Pivovarova, A. (eds.) *Reforming Child Welfare in the Post-Soviet Space: Institutional Change in Russia* (London & New York, Routledge), 181-197.

Skokova, Y., Paper, U. and Krasnopol'skaya I. (2018) 'The non-profit sector in today's Russia: between confrontation and co-optation', *Europe-Asia Studies 70*(4), pp. 531–563. doi: https://doi.org/10.1080/09668136.2018.1447089.

Stein, M. 2006. Young People Aging out of Care: The Poverty of Theory, *Children and Youth Services Review, 28*(4): 422–434.

Tarasenko, A. (2021) 'Institutional variety rather than the end of residential care: Regional responses to deinstitutionalisation reform in Russia', in Kulmala et al.

(eds.), *Reforming child welfare in the post-Soviet space: Institutional change in Russia*. London and New York: Routledge, pp. 95–117.

Törrönen, M., Munn-Giddings, C., Gavriel, C., O'Brien, N., and Byrne, P. (2018) *Reciprocal emotional relationships: Experiences of stability of young adults leaving care*. Publications of the Faculty of Social Sciences 75. Helsinki: University of Helsinki.

Viuhko, M. (2020) *Restricted agency, control and exploitation. Understanding the agency of trafficked persons in the 21st century Finland*. Helsinki: Heuni.

10 Youth wellbeing in "Atomic Towns"

The cases of Polyarnye Zori and Pyhäjoki

Ria-Maria Adams, Lukas Allemann and Veli-Pekka Tynkkynen

Introduction

In this chapter we present and analyse two case studies of how young people living in the "atomic towns" of Pyhäjoki (Finland) and Polyarnye Zori (Russia) perceive wellbeing. We look at the two cases through the lens of ethnographic methods and policy analysis to provide insights into how local youth are connected to the energy companies operating nuclear power plants in their hometowns. The two sites differ greatly in their socio-economic, cultural and political setting, as well as in the development stage of the respective power plants. Despite obvious differences, we have found striking similarities in young people's perceptions of what the "good life" means to them in these prospective/current nuclear towns.

Pyhäjoki and Polyarnye Zori may seem like an unusual pair for comparison, ostensibly sharing little beyond the fact that the former is, and the latter will likely become, dominated by the nuclear power industry. The common yet different connection of these towns to Rosatom offers insights into the relations between the local youth and a major industrial player in places that lack economic diversity. In Polyarnye Zori, Rosatom is a pervasive actor: it produces electricity—its core product—provides municipal housing maintenance and transportation and sees to the catering at large events. This kind of "town-forming company" (*gradoobrazuiushchee predpriiatie*) is a common phenomenon in Russian single-industry towns (see also other chapters in this volume: Bolotova; Ivanova et al.; Simakova et al.). At present, the power company in Pyhäjoki does not have a comparably dominant role, but the community is on track to become an "atomic town" as well. In reviewing and analysing our data from these contrasting field sites, we bring to light the similarities as well as the differences in the factors young people regard as important to their personal wellbeing.

We start by providing insights into our two field sites. Next, we outline our methods and embed our case studies in the relevant policy settings, these being energy politics and geopolitics. Then we compare and describe the studies, which are informed by and grounded in anthropological theories of wellbeing. In our analysis, we contribute to the understudied area of comparing towns reliant on the nuclear energy industry in disparate cultural and

DOI: 10.4324/9781003110019-14

policy settings in the Arctic. Petrov et al. (2017, p. 56) urge that research efforts should increasingly focus on

> understudied issue areas with global-national-regional-local linkages, in order to better understand outlooks and pathways for Arctic sustainable development as well as the Arctic's role in global processes and sustainable development challenges.

Finally, we discuss how the policy analysis and the ethnographic findings can help to understand young people's decisions about their future in the two regions studied.

Polyarnye Zori and Pyhäjoki: two contrasting field sites?

Pyhäjoki is a municipality located in North Ostrobothnia, some one hundred kilometres south of Oulu, the most populous city in northern Finland. Pyhäjoki, the central town in the municipality, is currently entering a new stage of development: the nuclear power plant will be built in an existing community and the population is expected to grow in the upcoming years as the power plant project progresses. The town has 3,146 inhabitants at the time of writing, with this population currently declining at a rate of 1.3 per cent annually (Statistics Finland 2019). The unemployment rate in 2019 was 10.1 per cent, which was slightly lower than Finland's general average of 11.3 per cent (Statistics Finland 2019). The municipal council of Pyhäjoki has developed an operational plan for creating new employment and business opportunities by promoting the area to newcomers and existing businesses alike (Pyhäjoki Municipality 2020). Despite ongoing construction work at the power plant site, the municipality has not succeeded in growing the number of inhabitants. In fact, in the year 2019, the number of inhabitants decreased by 2.2 per cent due to outmigration (Statistics Finland 2019). The planning of Fennovoima's "Hanhikivi 1" project started back in 2007 and Fennovoima continues to wait for the final permits to build the nuclear power plant by Rosatom. Various factors, either within or outside the company, have slowed the progress of the construction work, these mainly relating to building permits and meeting the safety requirements of the Finnish Radiation and Nuclear Safety Authority (Fennovoima 2020).

Polyarnye Zori is a town in Murmansk Region, North-West Russia, about two hundred kilometres south of Murmansk, the largest city in the world above the Arctic Circle. According to the most recent census, Polyarnye Zori had a population of 14,389 (Rosstat 2019). With a population decline of 6.3 per cent over the past decade (Rosstat 2009; Rosstat 2019), the outmigration rate has been lower than in most other towns of Murmansk Region. The current rate of registered unemployment is also low at 3.2 per cent (PZ City 2020). The single most important employment sector in the town is the nuclear power plant (owned by Rosenergoatom (Rosatom's civilian subsidiary)) and its subcontractors. The main subcontractor for maintenance is Kolatomenergoremont (owned by Rosenergoatom), but there are also private subcontractors. The town was built in the early 1970s, at the same time as the power plant, meaning that the creation of the town was tied to the creation of the power plant. This dependence is still pervasive but since the end of the Soviet Union private small- and medium-scale entrepreneurship have formed a secondary socio-economic pillar, as in any other Russian single-industry town. Rosatom's structures dominate not only large parts of the residents' professional lives but also much of their leisure time. To cite but a few instances of Rosatom's omnipresence, the town park, the stadium and the sports centre are sponsored by the company; apartment block maintenance is provided by one of its subsidiaries; and the town's favourite *pirozhki* (stuffed buns) come from the power plant's bakery.

For the people of Pyhäjoki, the coming of the power plant entails multiple ramifications, one being the hope that the resulting growth in population will add a variety of services. The municipality, as Strauss-Mazzullo (2020, p. 38) notes, has become a place on the map beyond the national borders of Finland. Polyarnye Zori has gone through the anticipated stages of development a long time ago and is now in the phase where the effects of a running power plant are visible in daily life. The area of North Ostrobothnia is known for having many adherents of the Lutheran religious movement known as Conservative Laestadianism (*vanhoillislestadiolaisuus*); members are particularly known for their active engagement in entrepreneurship, especially in the construction industry (Linjakumpu 2018). This might be part of the reason why the municipality voiced a strong interest in obtaining the nuclear megaproject and thus ensuring prospective employment and continuous revenue for the community members (Linjakumpu 2018; Strauss-Mazzullo 2020, p. 28). While this article does not deal with the effects of religious movements, it is important to mention this specific characteristic since it influences employment in the area and helps to understand why such a megaproject is being constructed in this particular place.

If one visits Pyhäjoki, Fennovoima (and thus indirectly Rosatom) is visible as a single, small, inconspicuous office in the centre of the municipality; Rosatom's visual presence in Polyarnye Zori is ubiquitous. This visibility in Polyarnye Zori has its historical roots in the practice of establishing single-industrial towns, one dating from the Soviet policies of economic specialization. With the nuclear power plant traditionally being essentially the only industrial employer in the atomic towns, city and regional governments have imposed extensive economic and social responsibilities on the operators (cf. Collier 2011).

By describing these two contrasting field sites, and by embedding policy analysis in our study, we demonstrate the structural differences between the communities in size, stage of company development, and political and cultural context. These distinctions are important in order to understand the outcome of our comparative ethnographic analysis of the local youth's perceptions, wishes and aspirations in both locations.

Methodological considerations and research ethics

Although doing research in two strongly differing towns, we have used the same qualitative methods to gain insights. Anthropological fieldwork lies at the core of our methodological approach, which featured open-ended individual interviews, focus group interviews and media analysis (Clark 2011; Silverman 2013, 2014; Olivier de Sardan 2015). The holistic approach of anthropology is valuable as it takes all different aspects of human living into account and considers both actions and verbal accounts equally (Eriksen 2010; Crate 2011). In the context of our research, we sought to listen to young people and to let their views be heard.

Participatory approaches to youth research are often described in "all or nothing" terms, meaning that young people participate either as active researchers or as passive research objects (Heath and Walker 2011, p. 8). In our research we aimed at finding a middle ground where we actively engaged in youth activities. Through several fieldwork visits by Adams (Pyhäjoki) and Allemann (Polyarnye Zori) between August 2018 and August 2020, we engaged with young people in a way which did not render them passive research objects but rather invited their insights on various topics of their hometowns. While doing research among young people, we tried to be sensible about the issue of the unequal power relations between us as researchers and the youth as the focus of our work. Young research participants may lack the resources, social networks and knowledge of those conducting the research and thus be in a potentially vulnerable position (Cieslik 2003, p. 2). We were aware that we were dealing with sensitive topics, and throughout the research process we were open with our informants about how we would process and store the data. The identities of the young people sharing their life stories, ideas and thoughts with us were anonymized in order to protect them from any possible exposure. At the same time, our analysis enables us to exemplify how young people are connected to their nuclear towns, how they perceive the industries in their hometowns, what wishes and aspirations they have and what opportunities they see in their future.

We were guided by the conviction that the task of finding out about what hedonic wellbeing means for young people could only be fulfilled successfully by working *with* them. In such co-productive research, we see the interviewee as on a par with the scholar (cf. Denzin 2009, pp. 277–305; Allemann and Dudeck 2019). Our interlocutors are not just sources of raw data that we tap into and then interpret. Rather, it is primarily the interlocutors who actively reflect on their own lives while talking to us (Bornat 2010). Our task as researchers is to connect these reflections with each other and with our field observations on political and corporate actors. Thus, we see the interviews not as factual data but as *first-stage* interpretations, on which we build our *second-stage* scholarly interpretations and recommendations.

The Finnish Youth Act (Ministry of Education of Finland, 2017) defines persons up to the age of 29 years as youths, while in Russia at the time of the research the age range was between 14 and 30 years (Government of the

Russian Federation 2014, see discussion of the age-range in Ivanova et al., this volume). The selection of research participants was guided by these respective definitions.

Young people's experiences can offer exceptional insights into the operation and the character of institutions. Roberts (2003) argues that changes in and links and mismatches between institutions become apparent through youth research by giving voice to young people's experiences, which then lays a basis for broader academic debates (Roberts 2003, p. 15). Our research results suggest that young people's views on how a "good life" is constructed in a nuclear town is marginally connected to the industry itself but relies on other considerations, such as educational opportunities, getting work in one's own specialization, having places to "hang out", having access to a functioning infrastructure, being connected to services (such as leisure activities and health services), being connected to nature, having a feeling of safety and being close to social networks of family and friends.

Geopolitics, the nuclear sector and corporate social responsibility

Both Russia and Finland are nuclear power-friendly states in that nuclear power plays a central role in national energy policies: in Russia 18 per cent, and in Finland 25 per cent of electricity is produced by nuclear power plants (International Energy Agency 2018). Moreover, people's attitude toward nuclear energy is relatively positive (Wang and Kim 2018). What is more, the nuclear sector is also a central element in Finnish-Russian trade relations and foreign policy. In Finland, nearly half of all energy consumed is of Russian origin, two-thirds of all energy imported comes from Russia, and nearly all of Finland's fossil (around 80 per cent) and nuclear (varies from year to year between 40 to 70 per cent) fuel comes from Russia (Statistics Finland 2017). Thus, when it comes to energy, the relationship between Finland and Russia is tight, yet at the same time very asymmetric: Finland accounts for a small percentage of Russia's energy exports while energy imported from Russia makes up a large share of total energy imports in Finland. The dependency of Finland's energy sector on Russian hydrocarbons, as well as on its nuclear power technology and nuclear fuel exports, gives Russia political leverage vis-à-vis Finland. The fear of losing the economic benefits gained from consuming, refining and further selling energy of Russian origin have an impact on Finland's policy considerations (Tynkkynen 2016; Jääskeläinen et al. 2018). Thus, the Pyhäjoki project between the Russian state-owned nuclear corporation Rosatom and the Finnish private enterprise Fennovoima includes a foreign policy dimension, not least as the project is symbolically and economically important for Russia. On balance, the social programmes promoted as part of nuclear projects should be understood as intertwined with other (energy and foreign) policy issues.

The main institutional actor in this context is Rosatom, Russia's state-owned corporation which controls the civilian and military use of nuclear energy in both the internal and the export markets. Rosatom's position is unique, as the company does not have to produce a profit. Accordingly, it is better positioned to promote a wide range of policy objectives set by the state both domestically and internationally. In Russia, nuclear power is prioritized in relation to other energy sectors, and internationally Rosatom has the possibility to increase Russian influence through very attractive deals for constructing nuclear power plants and supplying uranium (Tynkkynen 2016). In the Finnish context, the role of Rosatom is exceptional, as the company will be the exclusive provider of uranium fuel to the Pyhäjoki plant for the first ten years. As long as the project progresses smoothly, the Pyhäjoki power plant, as cooperation in the area of nuclear power, officially stands to make a key contribution to enhancing good relations between Finland and Russia (Ministry of Foreign Affairs of Finland 2016; Putin and Niinistö 2017). If problems should occur during the project, they will reflect badly on relations between the countries.

How does this wider political context then affect social policies and the youth in atomic towns? The political context has very much to do with the corporate social responsibility (CSR) practices of Rosatom in Russia, and especially in Finland. CSR is a soft-power tool used by state actors, in addition to all other means, to further their objectives. Rosatom's CSR in Finland is important precisely for geopolitical and geo-economic reasons, and the community's youth are an important target of CSR efforts.

CSR is the instrument by which companies approach the issue of wellbeing. Following a global trend, but also replicating many practices from the Soviet era, CSR is an integral part of social policy among Russian energy giants like Rosatom (comp. Saxinger et al. 2016; Wilson and Istomin 2019). Rosatom promotes social policy under the heading of sustainability. CSR is implemented based on general objectives of the corporation addressing traditional social and work-related issues of the workers, but also promoting the wellbeing of workers' families and, in particular, their children. The youth are explicitly chosen as a focus group within Rosatom's CSR. Work on sustainability is operationalized through two special programmes "Rosatom School" and "Rosatom's Territory of Culture", which promote school children's skills in the natural sciences and nuclear physics, in particular, but also offer youth possibilities to engage in and enjoy music, arts and sports (Rosatom 2020a).

The role of Finland and Fennovoima's nuclear power plant as an important reference for Rosatom is also visible in the company's CSR activities. The webpage of Rosatom (2020b) bills the success of "Rosatom School" in the following terms:

> Since 2016, about 1,000 kids from 25 countries have taken part in nine International Smart Holidays with the "ROSATOM School" in Russia, Indonesia, Bulgaria, Hungary, Finland, Thailand and Turkey. [...] For example, in 2017–2018 schoolchildren from Russia and Finland implemented the international project "Educational Tourism". In this project

10 schoolchildren from Sarov visited Pyhäjoki and schoolchildren from Pyhäjoki visited Sarov.

Providing amenities under the heading of CSR by a powerful industrial actor may enhance the wellbeing (basic needs) of people in general and the youth in particular. However, at the same time CSR consolidates the company's political and economic power, potentially diminishing people's choices as to how to live a good life (extended and broader needs). Especially in a single-industry town there is a risk that decisions on which youth activities to develop or sponsor are guided by the dominating company's own interests. In a small town, this both broadens and limits people's choices: on the one hand, a powerful sponsor may make it possible to provide the youth opportunities that would otherwise be impossible, such as the sports school in Polyarnye Zori, which offers a wide variety of different sports. On the other hand, activities that do not coincide with the larger sphere of interests of the dominant sponsor may be dismissed despite a demand from the youth. In Polyarnye Zori, where the Rosatom company is already an established player, this is reflected in an ambivalent attitude towards the opportunities that the company offers (Allemann and Dudeck 2019).

Youth wellbeing: Eudaimonic and hedonic perceptions

What does "a good life" mean for young people living in very different countries and circumstances and what are their visions of wellbeing? Fischer (2014) and numerous other theorists on wellbeing (see discussion in Stammler and Toivanen, this volume) argue that wellbeing, across cultures, cannot be reduced to material conditions alone, which is in line with our findings in Pyhäjoki and Polyarnye Zori. Perceptions of wellbeing from the young people's point of view in the two atomic towns entail more than having access to work or a stable income. Young people long for functioning social networks, mobility and a choice of educational opportunities.

Lambek (2008) addresses the importance of wellbeing in human sciences, while simultaneously acknowledging that it is a 'problematic' topic in terms of "measuring or bringing about other people's wellbeing" (Lambek 2008, p. 115). He argues that "measuring from the outside someone else's quality of life" (p. 116) might seem inconceivable, whereas through their ethnographic research anthropologists can acquire more perspectives "from inside" and thus make valuable contributions to theorizing wellbeing, as the introduction to this volume shows (Stammler and Toivanen, this volume). Moreover, Lambek points out that ethnography must suffice as the basis for more general claims concerning an anthropological perspective on wellbeing by elaborating how wellbeing is constituted:

> Well-being does not occur in the abstract. As human life is culturally constituted, so well-being only makes sense with respect to the contours of a particular way of life; particular structures of persons, relations,

feeling, place, cosmos, work and leisure. Another way of saying this is that quality of life cannot be simply open freedom of choice. Well-being must include guides and orientations in the making of choice or the exercise of judgement, ones that affirm people's intuitions.

(Lambek 2008, p. 125)

Our analysis shows that wellbeing has different meanings for young people. While some proportion of the youth perceived their hometown as a place that lacks activities, others described exactly the opposite, as for them their hometowns provide everything that they need.

A range of scholars focusing on wellbeing distinguish between hedonic happiness and eudaimonic happiness (Ryan and Deci 2001; Mathews and Izquierdo 2009; Fischer 2014; Edwards et al. 2016; Johnson et al. 2018). Hedonic happiness refers to everyday, short-term contentment, such as buying a long-desired item or satisfying "a man's own desire to play instead of working" (Hobsbawm 1968, p. 85). Eudaimonic happiness denotes a broader, overall life satisfaction. While hedonic happiness is more ephemeral, eudaimonic happiness as well is far from being a static condition or a final goal. As Fischer notes, "good life is not a state to be obtained but an ongoing aspiration for something better that gives meaning to life's pursuits" (2014, p. 2). Eudaimonic happiness consists of many trade-offs, often at the cost of hedonic happiness. The pair "can well be at odds with each other, a tension familiar to most from daily life" (Fischer 2014, p. 2).

In this research we combine these two notions of happiness as key concepts for a holistic understanding of wellbeing. People pursue their individual visions of a good life, but the concept of wellbeing is morally laden with ideas about value, worth, virtue and what is good or bad, right or wrong (Fischer 2014, pp. 4–5). Adding to this observation, Fischer argues that "adequate material resources, physical health and safety, and family and social relations are all core and necessary elements of wellbeing" (p. 5). This means that aspirations are limited by the capacity to aspire. Such constraints on aspiration and agency may be social norms, legal regulations or the labour market. The individual's will is important, but there also has to be a way (Appadurai 2013, pp. 179–195; Fischer 2014, p. 6).

In our analysis, we sought to achieve a cross-cultural comparison that informs us about the ways in which "beliefs, practices and institutions impinge on happiness" (Thin 2008, p. 135). Thin argues that ethnographic methods and analytical approaches enable researchers "to observe and discuss the quality of human experiences, the ways people feel about their lives in general and about specific institutions and practices in particular" (p. 135). There is a rather limited corpus of anthropological literature developing a systematic interest in the subjective, experiential aspects of wellbeing (see Ortner 2016, and the introduction, this volume for an overview). The topic is mostly associated with and dominated by psychological or economic perspectives (see Johnson et al. 2018 for a recent overview). Brown (2013) attempts a comparison between the "atomic cities" of Ozersk (Russia) and

Richland (United States) and the impact of families. However, our aim is not merely to compare our case towns, but rather to demonstrate young people's views on a good life in their respective hometowns.

Based on these considerations, we ask: What are the important factors prompting young people to live in, stay in or return to Pyhäjoki and Polyarnye Zori? What does wellbeing mean to them? Our results suggest that wellbeing from the youth's perspective strongly emanates from aspects of focusing on the self, the present moment and consuming what a young person wants or needs. This, essentially, is a hedonic perception of wellbeing.

The meaning of good life for young people in nuclear towns

The majority of our informants addressed the need and desire to have places where young people can "hang out" together. In Pyhäjoki the municipality runs a youth centre which is open on weekend nights for youth to informally spend time together in a safe space supervised by a youth worker. A billiard table, game consoles, board games, TV, a stereo, sofas and a kitchenette for drinks and snacks are central elements of the centre, which is used primarily by youths under 18 years. In addition, the local church has a supervised youth space once a week, which attracts many young people (under 18). One collaborative project between the church and the municipality is a small workshop (*moottoripaja*), where young people can come and fix their motorbikes once a week under the supervision of volunteers (usually older local men). Another, easily accessible place in the centre of the municipality is an old youth association house (*nuorisoseuratalo*), which local young people and activists have turned into an indoor skating rink. According to our informants, the facility has a long history of attracting young people from outside Pyhäjoki for events and parties but is now rather run down; young people can hang out there or skate at their own risk. However, many parents do not allow their teenagers to go to the indoor rink, because it is not supervised. Nevertheless, younger youth value the rink, precisely because they are not supervised there, which gives them the feeling of freedom to do what they want. Other places offering possibilities to meet socially are private cottages or homes, where groups of friends are likely to spend their free time. In addition, the local upper secondary school offers its students space to hang out on the school premises. The newly re-opened and only local pub (*Dado*) has been well received, especially among youth aged over 18, who for some years had no place to gather evenings and weekends. In sum, sufficient availability of places where they can spend leisure time is an important factor of wellbeing for young people in Pyhäjoki, as this female teenager indicated:

> If I could wish something for our town it would be a burger restaurant. On the main road, just beside the shop would be a good place. We could go there, get some food or drinks and just hang out with our friends. It would be nice to have clothing stores as well, but I know that it is not a realistic wish because of the size of our town.

Having access to a variety of free-time activities is essential for keeping youth in their hometown. Pyhäjoki offers a range of activities and facilities: ice hockey, tennis, hunting, motocross tracks, swimming, floor hockey, soccer, frisbee golf, cross-country skiing tracks, a gym and running tracks, to name the most common ones. However, if young people want more specialized facilities like an indoor pool or horseback riding, they have to travel to the neighbouring towns. In Pyhäjoki different organizations and sport clubs can apply for financial support from Fennovoima; according to informants, however, the money granted is usually not decisive for keeping the organizations running. In Finland leisure activities (*harrastustoiminta*) are considered a core element of youth wellbeing, and therefore access to various activities is guaranteed by the Youth Act (Ministry of Education of Finland, 2017). Youth services, which in Finland include professional youth workers, are provided mainly by the state, not by private actors (European Commission 2017).

In Polyarnye Zori, the municipality offers young people a wide range of organized leisure activities, not least thanks to the presence of Rosatom as a powerful sponsor. There is a sports school, an arts and music school and a wealth of creative activities at the House of Creativity and the House of Cultures, examples being dance classes, theatre groups or scale-model building. All of these activities are free of charge, but they require regular attendance and thus a certain level of commitment. Another option is the Club of Interesting Things (*Klub interesnykh del, KID*), also funded by the municipality and free of charge, but less structured. There is only one adult supervisor present and the idea is that kids and teenagers teach each other useful and fun things, such as playing the guitar, playing ping-pong or writing a convincing speech. What all these places have in common is that they require a young person to be active. If someone wants just a place to "hang out", there are only commercial venues—shopping centres, restaurants and coffee shops—or socially marginalized options—the railway station or an abandoned construction site. The first category is essentially a limited range of cafés. For teenagers specifically, there is what is known as a "time café" (*Lemonade*), where youngsters pay for the time spent there but do not need to buy anything. It fulfils the function of a private youth centre; teenagers socialize, play games or just relax in a cosy atmosphere, supervised by the owner, a young man who maintains an easy-going relationship with his returning customers. In atmosphere the place comes closest to a Finnish youth centre, but with the big difference that it is not free of charge. This automatically excludes the young people from less wealthy families, but also those who (or whose parents) prefer to spend their money on something other than paying for the time spent in a place. The café-goers largely reflect the prevailing societal rift in the town created by the big gap in income between the "powerplanters" (*stantsionniki*) and the "non-powerplanters" (*nestantsionniki*), a difference sometimes described by the latter, less privileged group, as creating two "castes". A young female illustrates the situation:

> Some parents simply cannot afford that [time café] or they don't allow [their children] to go there. [...] For instance, some friends may say 'let's hang out there' and others say 'I don't have the money', and so in the end we all end up at the railway station or the petrol station.

Thus, for many youngsters the only relatively attractive alternative is free "hanging out" in non-supervised public spaces. In summer, such places are the nearby woods and the backyard of the cinema; in winter, it is mainly the waiting room of the local railway station. Asked in a group discussion why they like to hang out at the railway station, a female teenager answered: "They don't chase us away from there. They don't ask for money; you can just sit there with your smartphone, and it's warm. That makes me happy."

Our comparison shows that in both Polyarnye Zori and Pyhäjoki young people would like to have more places where they could meet up in their hometown:

> This skating place is like a second home for me. I come here almost every day after school. I hang out here with my friends because there is no other place where we can go and just be. Sometimes we skate, sometimes we just sit around and talk. Nothing special. Sometimes we go to the shop to get snacks and drinks and then we return.
>
> (young male, Pyhäjoki)

In Pyhäjoki young people are generally content with what has been provided, given the size of the municipality and the relative proximity of the bigger municipalities of Raahe, Kalajoki and Oulu. The nature surrounding the community has been mentioned in particular in many conversations as an important and empowering place of regeneration. Decision-makers in the municipality cite the possibility of carrying out young people's wishes once actual construction on the nuclear power plant starts and the town grows. In the meantime, local youth keep using the available services and continue to visit nearby towns on a frequent basis to meet their needs. The same applies in Polyarnye Zori. Both the surrounding nature and other urban centres, such as nearby Kandalaksha and even more distant Murmansk, offer opportunities to get away and enjoy a change. In both Pyhäjoki and Polyarnye Zori young people say they would like a place offered by the municipality where they could just "hang out" without necessarily having to pursue goal-oriented, constructive activities.

In both of our case sites, access to education becomes an issue after the level of basic school. In Pyhäjoki, the municipality offers upper secondary education (*lukio*) with a focus on entrepreneurship, which is a popular choice among local youth and also attracts youth from the neighbouring towns. However, after graduation, at the latest, young people need to move elsewhere if they wish to continue their studies. For any vocational training young people have to leave town and they will either have to put up with commuting long distances to school, inconvenienced by the inadequate public transportation, or face having to move from home to another town at a relatively young age (around 15 or 16 years). Decisions on what to study tend to be taken on a very individual level of interest and these choices are rarely linked to the arrival of an industry in town. At the time of our research there were no specific industry programmes in place designed to train young people for professions in the power plant, yet educating local youth to work in the plant can be understood as a facet of CSR. However, the vocational schools in the region (Koulutuskeskus Brahe 2020) provide a variety of educational opportunities, which include qualifications in technical fields and construction.

Polyarnye Zori has one educational institution at the post-secondary level: a vocational school with several degree programmes to choose from. Originally, the school was conceived as a supplier of an educated workforce for the nuclear power plant. Today, as a result of the regional budgeting based on competition between educational institutions, the school is trying to diversify its educational offerings and also to attract students from outside of Polyarnye Zori. About half of the students come from other cities. However, attracting them is not easy and the school expends considerable effort in doing so. About half of the curricula are designed especially for the energy sector. Two-thirds of the students are male, one-third female, usually starting their training around the age of 15. Recently, the school introduced a new curriculum serving the hotel trade, diversifying its offerings to perhaps attract more female students and thus reduce the gender imbalance in the student body. The school has limited co-operation with the nuclear power plant. This centres on obtaining internships for students and does not include any employment programmes. The situation differs therefore considerably from that of other Russian Arctic single-industry towns, such as neighbouring Kirovsk and Apatity, Neryungri and Novyi Urengoy, where the dominant industry in town is heavily involved in educating specialists on the post-secondary level and offers employment for many graduates (see Simakova et al., this volume).

As a result, the employment situation for young people in Polyarnye Zori is not an easy one. Information gathered in interviews with the vocational school headmaster, pupils and power plant employees clearly indicates that the school produces far too many specialists for the nuclear sector. Students consistently report that it is very difficult to find a good job in Polyarnye Zori despite Rosatom being widely perceived as one of the most stable and well-paying employers in the entire region. According to the headmaster, only about 10 per cent of the young people graduating with a vocational education get a job at Rosatom. Local family ties play a significant role when it comes to getting a job with no more than a vocational-level education, with dynasties of power plant workers being a common phenomenon.

The nuclear energy sector also has an above-average need for specialists with a higher education, mostly in engineering. Indeed, about 40 per cent of the vocational school's graduates leave Polyarnye Zori to go on to complete a higher education. However, informants have repeatedly claimed that the proportion of locals among the power plant's recruited employees with a higher education is low. Thus, according to several informants, among employees with a higher education an overwhelming majority are not from Polyarnye Zori. There seem to be informal loyalty ties (Ledeneva 1998, 2006) between several cities in the Urals and Siberia, which have specialized universities and from where some of the management come. A young male power plant employee described the situation as follows:

> Yes, unfortunately it's like that. There are loads of those who come from those few universities. It's as if they deliver them here in buses. As a local I feel frustrated about this. [...] But if Mum and Dad work there then yes, chances are higher that you get in. Because they will trust you more.

On the other hand, there are plenty of job openings in medicine and in the school that cannot be filled because there are not enough specialists who want to move to Polyarnye Zori. The problem stems from the low salaries and lack of support programmes. For most of the local youth, what remains are less attractive jobs, such as being a salesperson or waiter, or moving away. A few become small entrepreneurs (see also Bolotova, this volume, on "forced entrepreneurship", pp. 53–76).

In Pyhäjoki finding suitable work is also a major concern for young people, as one male participant noted:

> There are certain sectors where you could get a job immediately if you wanted, like taking care of elderly people or working in some construction company. But that's not what I am interested in. If you want something else, you either have to commute or come up with creative solutions, like starting your own business. But sometimes you have no choice other than to move away if you want to find something matching your education.

Once young people have chosen an educational path and have graduated, it is important for them to find work that satisfies their expectations. This being the case, they tend to move elsewhere to pursue their career dreams and aspirations if they cannot find a suitable place in their municipality. However, there is a group of young people, especially young males, who express hope in being able to find a job in construction associated with the nuclear power plant. Not unlike their counterparts in Polyarnye Zori, many of our informants in Pyhäjoki assume that local labour will not be needed as much as external expertise and therefore hopes of getting employed by the power plant are not too high.

Interestingly, our analysis shows that environmental concerns in times of climate change do not seem to be relevant in the view of the northern youths in the two countries. It seems that young people are more concerned with the lack of services and infrastructure rather than with what the presence of a nuclear power plant will do to their environment. The nuclear power plant in Pyhäjoki is being marketed as "green energy", but this discourse understandably does not include local environmental issues related to nuclear power, such as the risk of accidents, nuclear waste management and so on. The green in this setting is the potential of nuclear power to cut national and global greenhouse gas emissions if it is to replace energy produced from fossil fuels. In Polyarnye Zori, where the city's identity is strongly linked to Rosatom, nuclear energy has a special presence in school tuition, and for most of the local youth ecological concerns about nuclear energy are not an issue. This uncritical attitude towards potential ecological hazards does not mean that Rosatom dominates young peoples' perspectives on their lives and aspirations. The opposite is true for many members of the older generations in Polyarnye Zori, who actively participated in the place's coming of age as a city.

In this section, we have focused on comparing four major aspects of wellbeing that local youth in both places identify as vital from their own perspectives: hangouts, leisure activities, educational possibilities and future work opportunities. These four elements open up a wide field of issues. Young people in both places have emphasized the importance of functioning relationships and family ties as key components of their wellbeing. Besides work and education, these are important factors that determine youth wellbeing and thus the motives for staying in or leaving a place. Through our ethnographic examples, we have demonstrated how a "good life" is constructed from the perspective of young people. The connection to nature and a desire for purposeful activities through "eudaimonic" leisure opportunities play an important role in the perceptions of wellbeing, but even more striking is the importance of friends and rather "hedonic" pastimes that is reflected in the wish for more places where one can simply "hang out". While the "eudaimonic" components are valued by our participants and promoted by the industries, we conclude that "hedonic" components of wellbeing should be given more attention when creating viable towns for young people.

Conclusion: wellbeing rewired

With this chapter we contribute to the discussion of wellbeing that is embedded in specific, national policy frameworks. According to Larsen and Petrov (2020, p. 80), the Arctic region faces significant challenges related to regional and local economic development, industrial production and large-scale resource extraction. The role of nuclear power in the Russian Arctic region has been important traditionally.

Young people in both Polyarnye Zori and Pyhäjoki tend not to associate their futures with the (potentially) strong presence of Rosatom/Fennovoima. The reason is perhaps that the employment opportunities for locals are perceived as rather scanty, or as not coinciding with their perception of a "dream job", despite the fact that in both countries the nuclear sector is seen as a stable employer. Working in the atomic sector is perceived as applying to a very specific occupational group and there are many other professions that appeal to young people. Income does contribute to achieving a satisfactory level of subjective wellbeing but, as Fischer suggests, it is not everything: "Increases in happiness level off dramatically after people reach a relatively low income threshold" (Fischer 2014, p. 8). Even if the salaries at the power plant in Polyarnye Zori are satisfactory, working there is not an attractive option for many young people because of the burdensome hierarchies and rigid workflows, similar to what Bolotova (this volume) found in Kirovsk. In Pyhäjoki, the final phase of building the actual power plant seems remote to for young people, as the process of obtaining permits and building has been going on for years with no concrete completion date in sight.

Our research confirms that income alone from possible employment in the local nuclear power plant is not enough reason or motivation for young people to stay in the towns of Pyhäjoki and Polyarnye Zori. Rather, there are various other factors, such as family ties and friends, as well as educational and employment paths, that lead to an individual decision to either stay, leave or return. Our research also shows that policy analysis is relevant for the overall outcome, as it highlights the structural frameworks in which the companies operate and provides answers, for example, to the question of why significant investments are being directed to social programmes and the youth in this exceptional sector.

We have shown needs common to the youth of both Polyarnye Zori and Pyhäjoki, such as places to hang out. A place for "purposeless" hanging out clearly is what the young people in Polyarnye Zori miss most. By contrast, the presence of a wide range of "educationally valuable" options to organize one's leisure time is seen as something provincial. Many informants look at the bigger cities, which offer more opportunities for freer and idler pastimes. For instance, some informants mentioned the municipal "Centre for youth initiatives" (Tsentr molodezhnykh initsiativ 2020) in the large city of Belgorod (southern Russia) as an example worth emulating. Similar to Finnish youth centres, this place offers a free and supervised space for spending time in ways that are not structured from above.

We suggest that hedonic happiness be taken more seriously as a factor for youth wellbeing. "Good life" scholars like Fischer (2014) tend to consider hedonic happiness as the less gratifying form of happiness because it is not long-lasting. However, they are presenting their arguments from the vantage point of a mature, adult person. In this perspective, wellbeing is framed in terms of permanent and stable levels, and these levels can only be achieved through eudaimonic happiness (purposeful activities). Implicitly, the same attitude can be identified among many administrators who organize teenagers' leisure time (an insight based mainly on observations from Polyarnye Zori). However, for young people, opportunities to fulfil one's needs for hedonic happiness seem to be a relevant factor for wellbeing and thus for staying in a single-industry town. Taking these wishes seriously implies creating more opportunities for short-term gratification, which is interesting to young people but which administrators and educators often dismiss as useless or even harmful. Such "hedonic", "useless" activities may include places to play computer games or to just "hang around" with peers. Based on direct interaction and interviews with youth, we argue that having enough opportunities to experience positive emotions from "hedonic" activities is very important for young people (especially teenagers), while ideas about "higher" goals, purposes and achievements have lower priority before one's personality is fully developed, that is, around the age of 20 (Johnson et al. 2018, p. 8). Therefore, it is important to differentiate between the wishes of younger youth (approximately those under the age of 20) and older youth, who may have already returned or are considering returning because they value the peace, safety, nature, family-friendliness and relatives around them. Taking these different needs seriously has the potential to make small towns more liveable for their young population, especially in encouraging youth to return once they have finished their education.

Furthermore, we have argued in this chapter that finding suitable work in our case towns is not always easy for young people. Either the required social networks are missing or the jobs available do not match the education and aspirations of young people. However, we also met a substantial number of young people who were satisfied and had found work in their fields of interest. In both Polyarnye Zori and Pyhäjoki these were mostly "older" youth (over 20 years), who had already been elsewhere and then returned to start a family or be close to their family and relatives. It is understandable and unavoidable that people may have educational aspirations that they cannot fulfil in their hometown and therefore move away. It is also unavoidable that a large proportion of these people will not come back. The factors that motivate an eventual return to one's hometown are ultimately determined by a wellbeing surplus compared to the life in a big city. Coming to appreciate this surplus has much to do with one's biographical path: experiencing the difficulties or drawbacks of life in the "big city" and mirroring them against experiences of happiness in the place of origin. By showing that wellbeing is constructed of many different layers that are embedded in a particular policy framework, we have sought to contribute to wellbeing theories in anthropology and beyond. Our research suggests that the municipalities of Arctic atomic towns, such as the ones we have analysed here, would do well to offer activities, facilities and opportunities, which fulfil the youth's needs for both, eudaimonic and hedonic happiness.

Acknowledgements

This publication was supported by the Finnish Academy (research project *Live, Work or Leave? Youth—wellbeing and the viability of (post)-extractive Arctic industrial cities in Finland and Russia*, 2018–2020, decision no. 314471) and by the University of Vienna (uni:docs fellowship programme).

References

Allemann, L. and Dudeck, S. (2019) 'Sharing oral history with Arctic Indigenous communities: Ethical implications of bringing back research results', *Qualitative Inquiry*, 25(9–10), pp. 890–906. doi: 10.1177/1077800417738800.

Appadurai, A. (2013) *The future as cultural fact: essays on the global condition.* London; New York: Verso, imprint of New Left Books.

Bolotova, A. (this volume) 'Leaving or staying? Youth agency and the livability of industrial towns in the Russian Arctic', in Stammler, F. and Toivanen, R. (eds) *Young people, wellbeing and placemaking in the Arctic.* London: Routledge, pp. 53–76.

Bornat, J. (2010) 'Remembering and reworking emotions: the reanalysis of emotion in an interview', *Oral History, 38*(2), pp. 43–52. doi: 10.2307/25802189.

Brown, K. L. (2013) *Plutopia: Nuclear families, atomic cities, and the great Soviet and American plutonium disasters.* Oxford: Oxford University Press.

Cieslik, M. (2003) 'Introduction: Contemporary youth research: Issues, controversies and dilemmas', in Bennett, B., Cieslik, M., and Miles, S. (eds) *Researching Youth.* London: Palgrave Macmillan, pp. 1–10.

Clark, C. D. (2011) *In a younger voice: Doing child-centered qualitative research.* New York (NY): Oxford University Press.

Collier, S. J. (2011) *Post-Soviet social: Neoliberalism, social modernity, biopolitics.* Princeton: Princeton University Press.

Crate, S. A. (2011) 'Climate and culture: Anthropology in the era of contemporary climate change', *Annual Review of Anthropology, 40*(1), pp. 175–194. doi: 10.1146/annurev.anthro.012809.104925.

Denzin, N. K. (2009) *Qualitative inquiry under fire: Toward a new paradigm dialogue.* Walnut Creek: Left Coast Press.

Edwards, G. A. S., Reid, L. and Hunter, C. (2016) 'Environmental justice, capabilities, and the theorization of well-being', *Progress in Human Geography, 40*(6), pp. 754–769. doi: 10.1177/0309132515620850.

Eriksen, T. H. (2010) *Small places, large issues: An introduction to social and cultural anthropology.* London: Pluto Press.

European Commission (2017) 'Youth policies in Finland – 2017'. *European Commission.* Available at: https://eacea.ec.europa.eu/national-policies/sites/youth-wiki/files/gdlfinland.pdf (Accessed: April 24 2020).

Fennovoima (2020) *Fennovoima, story of Fennovoima.* Available at: https://www.fennovoima.fi/en (Accessed: April 25 2020).

Fischer, E. F. (2014) *The good life: aspiration, dignity, and the anthropology of wellbeing.* Stanford: Stanford University Press.

Government of the Russian Federation (2014) *Osnovy gosudarstvennoi molodezhnoi politiki Rossiiskoi Federatsii na period do 2025 goda.* Available at: http://government.ru/docs/15965/ (Accessed: April 25 2020).

Heath, S. and Walker, C. (eds) (2011) *Innovations in youth research.* New York, NY: Palgrave Macmillan.

Hobsbawm, E. J. (1968) *Industry and empire.* Harmondsworth: Penguin (The Pelican Economic History of Britain).

International Energy Agency (2018) *World energy balances 2018.* OECD Publishing, Paris: OECD. Available at: https://www.oecd-ilibrary.org/energy/world-energy-balances-2018_world_energy_bal-2018-en;jsessionid=PH1blxB-fJRIRsqlLHqdaNlv.ip-10-240-5-96 (Accessed: April 25 2020).

Ivanova, A., Oglezneva, T. and Stammler, F. (this volume) 'Youth law, policies and their implementation in the Russian Arctic', in Stammler, F. and Toivanen, R. (eds) *Young people, wellbeing and placemaking in the Arctic.* London: Routledge, pp. 147–169.

Jääskeläinen, J., Höysniemi, S., Syri, S. and Tynkkynen, V-P. (2018). Finland's dependence on Russian energy – Mutually beneficial trade relations or an energy security threat? *Sustainability, 10*(10), pp. 1–25.

Johnson, S., Robertson, I. T. and Cooper, C. L. (2018) *Well-being: productivity and happiness at work*. 2nd ed. Cham, Switzerland: Palgrave Macmillan.

Koulutuskeskus Brahe. (2020) *Koulutuskeskus Brahe. Ainakin sata syytä tarttua unelmaasi!* Koulutuskeskus Brahe. https://www.brahe.fi/ (Accessed: December 18 2020).

Lambek, M. (2008) 'Measuring - or practising - well-being?', in Jimenez, A. C. (ed.) *Culture and well-being: Anthropological approaches to freedom and political ethics*. London: Pluto Press, pp. 115–133.

Larsen, J. N. and Petrov, A. N. (2020) 'The economy of the Arctic', in Coates, K. S. and Holroyd, C. (eds) *The Palgrave Handbook of Arctic policy and politics*. Cham, Switzerland: Palgrave Macmillan, pp. 79–95.

Ledeneva, A. V. (1998) *Russia's economy of favours: Blat, networking, and informal exchange*. Cambridge, UK; New York (NY): Cambridge University Press (Cambridge Russian, Soviet and post-Soviet studies, 102).

Ledeneva, A. V. (2006) *How Russia really works: The informal practices that shaped post-Soviet politics and business*. Ithaca: Cornell University Press.

Linjakumpu, A. (2018) *Vanhoillislestadiolaisuuden taloudelliset verkostot*. Tampere: Vastapaino.

Mathews, G. and Izquierdo, C. (eds) (2009) *Pursuits of happiness: Well-being in anthropological perspective*. New York: Berghahn Books.

Ministry of Education of Finland (2017) Finnish Youth Act. Available at: https://minedu.fi/documents/1410845/4276311/Youth+Act+2017/c9416321-15d7-4a32-b29a-314ce961bf06/Youth+Act+2017.pdf (Accessed: November 23 2020).

Ministry of Foreign Affairs of Finland (2016) '*Ulkoministeriön lausunto*'. Available at: https://tem.fi/documents/1410877/2616019/Ulkoministeri%C3%B6n+lausunto.pdf (Accessed July 27 2021).

Olivier de Sardan, J.-P. (2015) *Epistemology, fieldwork, and anthropology*. New York, NY: Palgrave Macmillan.

Ortner, S. B. (2016) 'Dark anthropology and its others: Theory since the eighties', *HAU: Journal of Ethnographic Theory*, 6(1), pp. 47–73. doi: 10.14318/hau6.1.004.

Petrov, A. N. et al. (2017) *Arctic sustainability research: Past, present and future*. New York, NY: Routledge.

Putin, V. and Niinistö, S. (2017) *Sovmestnaia press-konferentsiia s Prezidentom Finliandii Sauli Niinistë, Prezident Rossii*. Available at: http://kremlin.ru/events/president/news/55175 (Accessed: April 25 2020).

Pyhäjoki Municipality. (2020) *Work and entrepreneurial life*. Available at: https://www.pyhajoki.fi/en/welcome (Accessed: April 25 2020).

PZ City. (2020) *Informatsiia osnovnykh pokazatelei rynka truda g. Poliarnye Zori na 01.04.2020 g*. Available at: http://www.pz-city.ru/index.php/novosti/1968-ekspress-informatsiya-osnovnykh-pokazatelej-rynka-truda-g-polyarnye-zori-na-01-04-2020-g (Accessed: April 6 2020).

Roberts, K. (2003) 'Problems and priorities for the sociology of youth', in Bennett, A., Cieslik, M., and Miles, S. (eds) *Researching youth*. London: Palgrave Macmillan, pp. 13–28.

Rosatom (2020a) *Social (S)*. Available at: https://www.rosatom.ru/en/sustainability/social-s/.

Rosatom (2020b) Project :Rosatom School. Available at: https://www.rosatom.ru/en/sustainability/project-rosatom-school/.

Rosstat (2009) *Chislennost' naseleniia Rossiiskoi Federatsii po munitsipal'nym obrazovaniiam*. Federal'naia sluzhba gosudarstvennoi statistiki. Available at: http://www.gks.ru/bgd/regl/B09_109/IssWWW.exe/Stg/d01/tabl-21-09.xls.

Rosstat (2019) *Chislennost' naseleniia Rossiiskoi Federatsii po munitsipal'nym obra-zovaniiam.* Federal'naia sluzhba gosudarstvennoi statistiki. Available at: https://www.gks.ru/storage/mediabank/mun_obr2019.rar.

Ryan, R. M. and Deci, E. L. (2001) 'On happiness and human potentials: A review of research on hedonic and eudaimonic well-being', *Annual Review of Psychology*, *52*(1), pp. 141–166. doi: 10.1146/annurev.psych.52.1.141.

Saxinger, G., Petrov, A., Kuklina, V., Krasnostanova, N. and Carson, D. (2016) Boom back or blow back? Growth strategies in mono-industrial resource towns – "east" & "west", in: *Settlements at the edge: Remote human settlements in developed nations.* Taylor, A., Carson, D., Ensign, P., Huskey, L., Rasmussen, R. and Saxinger, G. (eds.), Cheltenham: Edward Elgar, pp. 49–74.

Silverman, D. (2013) *Doing qualitative research: A practical handbook.* London: SAGE.

Silverman, D. (2014) *Interpreting qualitative data.* London: SAGE.

Simakova, A., Pitukhina, M. and Ivanova, A. (this volume) Motives for migrating among youth in Russian Arctic industrial towns', in Stammler, F. and Toivanen, R. (eds) *Young people, wellbeing and placemaking in the Arctic.* London: Routledge, pp. 17–31.

Stammler, F. and Toivanen, R. (eds) (this volume) 'The quest for a good life: Contributions from the Arctic towards a theory of wellbeing', in Stammler, F. and Toivanen, R. (eds) *Young people, wellbeing and placemaking in the Arctic.* London: Routledge, pp. 1–13.

Statistics Finland (2017) '*Energian tuonti ja vienti alkuperämaittain*'. Available at: http://pxnet2.stat.fi/PXWeb/pxweb/fi/StatFin/StatFin_ene_ehk/statfin_ehk_pxt_004_fi.px/?rxid=51014 c45-bdab-4956-8cac-d90d86bd14a7. (Accessed: February 26 2020).

Statistics Finland (2019) *Pyhäjoki statistics. Tilastokeskus.* Available at: https://www.stat.fi/tup/alue/kuntienavainluvut.html#?active1=625&year=2020 (Accessed: April 25 2020).

Strauss-Mazzullo, H. (2020) 'Promise and threat', in Tennberg, M., Lempinen, H., and Pirnes, S. (eds) *Resources, social and cultural sustainabilities in the Arctic.* London: Routledge, pp. 27–40.

Thin, N. (2008) '"Realising the substance of their happiness": How anthropology forgot about homo gauisus', in Jimenez, A. C. (ed.) *Culture and well-being: Anthropological approaches to freedom and political ethics.* London: Pluto Press, pp. 134–155.

Tsentr molodezhnykh initsiativ (2020) Available at: http://cmi31.ru/ (Accessed: April 8 2020).

Tynkkynen, V.-P. (2016) 'Russia's nuclear power and Finland's foreign policy', *Russian Analytical Digest*, *11*(193), pp. 2–5. Available at: http://www.css.ethz.ch/content/dam/ethz/special-interest/gess/cis/center-for-securities-studies/pdfs/RAD193.pdf.

Wang, J. and Kim, S. (2018) Comparative analysis of public attitudes toward nuclear power energy across 27 European Countries by applying the multilevel model, *Sustainability*, *10*(5), 1518. doi: 10.3390/su10051518.

Wilson, E. and Istomin, K. (2019) Beads and trinkets? Stakeholder perspectives on benefit-sharing and corporate responsibility in a Russian oil province, *Europe-Asia Studies*, *71*(8), pp. 1285–1313. doi: 10.1080/09668136.2019.1641585.

Index

Page numbers in *Italics* refer to figures; **bold** refer to tables and page numbers followed by 'n' refers to notes numbers

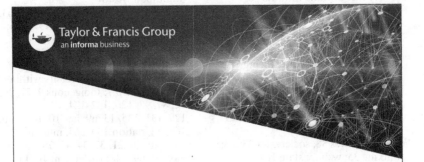

Printed in the United States
by Baker & Taylor Publisher Services

Printed in the United States
by Baker & Taylor Publisher Services